GSM: Evolution towards 3rd Generation Systems

GSM

Evolution towards 3rd Generation Systems

Edited by

Zoran Zvonar
Analog Devices, Wilmington, Massachusetts, U.S.A.

Peter Jung
Siemens AG, Munich, Germany

and

Karl Kammerlander
Siemens AG, Munich, Germany

KLUWER ACADEMIC PUBLISHERS
BOSTON / DORDRECHT / LONDON

A C.I.P. Catalogue record for this book is available from the Library of Congress.

ISBN 0-7923-8351-6

Published by Kluwer Academic Publishers,
P.O. Box 17, 3300 AA Dordrecht, The Netherlands.

Sold and distributed in North, Central and South America
by Kluwer Academic Publishers,
101 Philip Drive, Norwell, MA 02061, U.S.A.

In all other countries, sold and distributed
by Kluwer Academic Publishers,
P.O. Box 322, 3300 AH Dordrecht, The Netherlands.

Printed on acid-free paper

Printed in the Netherlands.

Contents

Preface

The Global System for Mobile Communication (GSM) is undoubtedly the most successful second generation digital mobile radio system. One of the key factors for this exceptional performance is the constant evolution of the GSM systems and its derivatives DCS-1800 and PCS-1900. The objective of the book is to address new concepts in the GSM system, dealing with both standardised features as well as theoretically and technologically feasible improvements, which contribute to evolutionary changes in general. Dynamic evolution of GSM presents a platform for the Universal Mobile Telecommunication System (UMTS) introduction and major trends in UMTS development will be addressed in this book, in particular progress towards new radio interface.

The book is the collection of individual contributions from a number of authors. The intention of the editors was to gather the most prominent specialist in the GSM area.

Ongoing efforts in GSM standardisation are focused on Phase 2+ with items planned to be added gradually on demand. Being mostly independent of each other, each of them could be introduced with little impact on the rest of the system. Speech remains the prime service of cellular systems with ongoing work on spectral efficiency taking into account the trade-off between cost of the system and transmission quality. Most notable result from novel speech service implementations is the introduction of new Enhanced Full Rate (EFR) speech coder. International railway organisation (Union Internationale des Chemins de Fer - UIC) has influenced the development of Advanced Speech Call Items (ASCI). Important novelties are group call service, voice broadcast service, enhanced multi-level priority and pre-emption scheme. The need for packet services has been covered under General Packet Radio Service (GPRS). High-speed data services have been increasingly important and the possibility of offering data channels of up to 64 kbits/sec is studied in the High-Speed Circuit Switched Data (HCSD) work item. Initiatives under CAMEL will cover intelligent networking (IN) within GSM framework.

Dual mode operation includes not only the dual-band solutions but also the inter-networking with other systems like DECT or satellite systems such as Globalstar, Iridium, and ICO. Also number of specific wireless applications like in building communications, WLL etc., have been developed based on GSM solutions. To support the required changes a number of issues have to be solved in the radio part of the system. Employment of adaptive antenna arrays for enhanced quality, interference suppression techniques and slow frequency hopping are among techniques to be presented in the GSM framework.

UMTS presents a path from existing second generation digital systems to the true third generation system that will allow wireless access for anyone, anywhere and at any time to broadband networks, truly integrating existing, conceivable and unanticipated new services, by providing seamless mobility. One of the key factors of UMTS is the radio evaluation which should provide broadband wireless reception and transmission of voice, data and images with data rates up to 2 Mbit/s. Presently, major efforts are directed towards the standardisation of the UMTS air interface in ETSI. Important contributors are pan-European projects such as ATDMA, which was carried out under the RACE II framework, and the ACTS project AC090 FRAMES. Major trends in UMTS multiple-access will be covered and future directions outlined.

The editors of the book would like to thank all authors for their contributions. The outlined material should cover the most important advances in GSM that will have lasting impact on the system influencing third generation wireless systems.

Editors

Chapter 1

The Status and Development of the GSM Specifications

An Overview

Friedhelm Hillebrand
Chairman of ETSI SMG

Key words: GSM, UMTS, mobile communication, TDMA, CDMA, mobile networks, mobile services, GSM Phase 2, GSM phase 2 +, GSM Release '96, '97, '98, ETSI SMG.

Abstract: GSM, the Global System for Mobile Communication is the world market leader for digital mobile communication systems with 62 % market share and more than 70 million users in 108 countries. The GSM success is caused by a number of GSM features: comprehensive services, system features, high quality, capacity and security. In addition the equipment cost for terminals and networks is low due to vendors' choice and market volume. A key factor for the success were the GSM specifications (and standards), since they provided a common, comprehensive, complete and stable system specification. The ongoing specification work for the evolution of GSM aims at supporting substantial growth and the exploitation of new market segments. In addition it is indispensable to maintain the integrity of the GSM platform in 900, 1800 and 1900 MHz. The content of GSM phase 2, the present working solution is described. The GSM Phase 2 + program brings GSM from second generation half the way to the third generation. The three GSM Releases '96,97 and 98 provide new services, new service creation tools, new terminal and network features like high speed data, new speech codecs, the use of Java like concepts in terminals and on smart cards (SIM). An outlook on the evolution towards Third Generation is given.

1. GSM'S ACCEPTANCE IN THE WORLD MARKET

1.1 GSM Specifications enable the global success of GSM

GSM the Global system for mobile Communication, is the world market leader for digital mobile communication systems with 62% market share, more than 70 million users and 251 networks in operation in 108 countries.

A wide range of manufacturers for SIM (subscriber identity modules), terminals, network systems and test systems supports this growth. Every SIM can work in every terminal. Every terminal can work in every network. Every network can work with every network to support roaming of terminals.

A key enabler for this success story were the GSM specifications (and standards), since they provided a common comprehensive complete and stable system specification. The ongoing evolution of the GSM specifications aims at supporting substantial growth and the exploitation of new market segments. In addition it is indispensable to maintain the integrity of the GSM platform.

1.2 GSM's Success Figures

1.2.1 GSM's World Market Share for digital Mobile Communication Systems (End 97)

There is a strong competition in the world market for digital mobile communication systems. GSM (Global System for Mobile Communication) is the market leader with a 62.2% share in the subscriber numbers. This means 70.2 million subscribers. GSM is based on an advanced TDMA (time division multiple access) radio technology. The Japanese PDC system (personal digital communication) - based on a basic TDMA radio technology - was able to achieve 24.1% or 26.8 million subscribers. The ANSI 54/136 (American National Standards Institute) TDMA system achieved 6.2%. The ANSI 95 CDMA (code division multiple access system) achieved 6.4%. This shows a clear leadership for GSM, and for TDMA technology.

Table 1. World market share: source for user numbers EMC (European Mobile Communication).

Standards	Radio Technologies	Users/million	Standards completed	%
GSM (900,1800,1900 MHz)	advanced TDMA	70.2	1991	62.2
PDC	basic TDMA	26.8	1992	24.1
ANSI 54/136 (800, 1900 MHz)	basic TDMA	6.9	1991	6.2
ANSI 95 (800,1900MHz)	narrow band CDMA	7.1	1991	6.4
Total		**110.0**		**100.0**

1.2.2 GSM User Numbers worldwide

GSM networks were introduced in 1991/2. Growth happened after an acceptable radio coverage was achieved and sufficient volumes of low cost, light weight terminals were available. There is still a very high relative and absolute growth.

Table 2. GSM users worldwide: source EMC and GSM MoU Association.

Date	GSM users worldwide/million
end 92	0.25
end 93	1.40
end 94	4.50
end 95	12.50
end 96	32.00
end 97	70.20
forecast for the end of 1999	160.00
forecast for the end of 2001	300.00

The GSM MoU Association has 293 network operators and regulators from 120 countries as members (end of April 98). This shows considerable short term growth potential for GSM networks and countries on air.

1.2.3 GSM Networks on Air

GSM networks went on air in the European Union first. The technology was accepted soon in the Arab World. It played a major role in the development of Middle/Eastern Europe (e.g. Hungary, Poland, Czech Republic).

It was accepted in Asia Pacific widely. Hong Kong, Australia and Singapore were forerunners. China Telecom is the biggest GSM network operator is in the world (8 million users).

In Africa it became a standard all over the continent. The highest subscriber numbers were achieved in South Africa. There it plays also a major role in the development of the black townships, since the network operators provide tens of thousands of payphones on their GSM networks.

In 1995 the first North American GSM 1900 network went into operation. In 1997 South America followed.

Today GSM is in operation in 108 counties in all continents. Since the GSM MoU Association has already members from 120 countries a further rapid growth of these numbers can be expected.

Table 3. GSM networks on air: source GSM MoU Association.

Date	Countries/areas on air	Networks on air
end 92	7	13
end 93	18	33
end 94	42	68
end 95	68	115
end 96	97	184
end 97	106	220
end of April 98	108	251

1.2.4 Mobile Satellite Services Operators (MSS) using the GSM Platform

The following MSS operators use the GSM platform in their networks and are committed to implement dual mode of operation and roaming with GSM networks based on dual mode handsets:

- ACeS (Asian Cellular Satellite Service)
- Globalstar
- ICO
- Iridium
- Thuraya

1.2.5 Manufacturers of GSM Equipment

GSM SIM (Subscriber Identity Modules) can be obtained from five major manufacturers.

Users have a very wide choice of terminals from a very competitive terminal market offering low prices due to high volume and intensive competition. Several other Japanese manufacturers announced plans for GSM terminal production (Matsushita, Nippon Penso). Users have a choice of several network operators in nearly all countries.

GSM network operators may choose network systems or subsystems from different vendors and maintain competition of their suppliers within their network.

There is a wide choice of antenna systems and repeaters. A very wide range of test systems for mobile stations, base station systems and the network subsystem is available.

Table 4. Major manufacturers of GSM equipment.

GSM SIM	GSM Terminals	GSM Network Systems	GSM Test Systems
Gem Plus	Alcatel	Alcatel	Alcatel
Giesecke & Devrient	Ericsson	Ericsson	Anite
Orga	Mitsubishi	Lucent Technologies	Hewlett Packard
De la Rue	Motorola	Motorola	Racal
Schlumberger	NEC	Nokia	Rhode & Schwarz
	Nokia	Northern Telecom	Schlumberger
	Panasonic	Siemens	
	Philips		
	Siemens		
	Sony		

1.3 GSM's Success Factors

This great success in the world market was caused by a range of GSM features:

Comprehensive Services and System Features

- GSM offers telephony, short message, fax and data services and a very comprehensive range of supplementary services.
- GSM offers network operators a choice of speech coding methods, full rate, dual rate (half and full rate) and enhanced full rate in order to meet their markets requirements in terms of capacity and quality.
- GSM allows global roaming between more than 100 countries including USA and Canada. Interstandard roaming GSM/AMPS and GSM/PDC provides a bridge to/from US AMPS networks resp. to/from Japan.

High Quality, Capacity and Security

- GSM offers high quality telephony, data and fax services.
- GSM offers high spectrum efficiency due to advanced TDMA technology with adaptive power control, slow frequency hopping and discontinuous transmission.
- GSM's advanced security and anti-fraud measures secure the customers privacy and the operators revenue.

Low Equipment Cost due to Vendor's Choice and Market Volume

- GSM is an open system standard, not dominated by the intellectual property rights of a single manufacturer.
- GSM offers the widest choice of terminal, network system and test system manufacturers.
- GSM equipment has a very high market volume and therefore very attractive prices.

2. GSM'S MEDIUM TERM EVOLUTION TOWARDS GENERATION 2.5

2.1 Definition of GSM by a joint Effort of ETSI and GSM MoU Association

The Global System for Mobile Communication is defined by two sets of documents:

- GSM specifications produced by ETSI (European Telecommunication Standards Institute) for GSM900 and 1800 and by ANSI for GSM1900. Those specifications which are needed for regulatory purposes (e.g. mobile station type approval) are converted into standards by ANSI or ETSI.

- GSM MoU Permanent Reference Documents (PRD) produced by the GSM MoU Association. They cover operational and commercial aspects and are not treated further in this article.

The GSM specifications and standards cover the technical aspects of GSM:

- services aspects
- network architecture
- selected interfaces
  - SIM/ME interface (ME = mobile equipment)
  - radio interface
  - network to network interface (MAP = mobile application part)
  - base station system to switching system interface (A)
  - base transceiver station to base station controller (Abis)

These GSM specifications ensure the compatible operation of all mobile stations in all networks and roaming between all networks. They allow the development of a mass market for mobile stations (world production was more than 50 million in 97). The GSM specifications form a stable basis for the development of network components and a volume production.

2.2 GSM Phase 2 Specifications, the Cornerstone for the Future

The GSM specification phase 1 were the basis for the opening of service in 1992. Phase 1 specifications were closed in March 95, i.e. no maintenance or upgrading is done any more.

Practically all present networks are based on phase 2 specifications, which were frozen in October 95. One year later all conditions had been settled for GSM phase 2 mobile stations to come to the market. The main additional features and functions of the full phase 2 standard are the enhanced compatibility mechanisms of the radio subsystem, allowing compatible evolution in the future, improvements of the radio link control (resulting in better spectrum usage), full specification of multi-band operation (allowing in particular dual band GSM 900/DCS 1800 mobile stations), improvements of data services (accommodation of short message services for all languages world wide) and changes in the network selection procedures of the mobile station.

GSM Phase 2: The Cornerstone for the Future

Phase 2 services:

- fax (group 3)
- improved and new supplementary services line identification services
  - call waiting
  - call hold
  - multi party
  - advice of charge
  - line identification

Optional phase 2 features:

- Half rate speech codec
- Enhanced full rate speech codec
- multi-band operation
- Unicode (providing coding for alphabets/character sets of all languages world-wide)
- Hierarchical cells

Phase 2 functions mandatory in the mobile:

- improved call selection, handover, frequency re-definition and radio link supervision, providing the basis for advance cell planning
- Phase 2 compatibility mechanisms and error handling
- Second cipher algorithm

The full phase 2 has - in addition to the new services - hooks for evolution and functions for more efficient spectrum usage and advanced cell planning.

2.3 GSM Phase 2+, the Program for a fast and modular Evolution

2.3.1 The Concept of Phase 2+ and the annual Releases

After the completion of the phase 2 specifications it was realized that the production of a complete specification phase was too inflexible. All items needed to be completed at the same time. A single item which was not ready could either lead to a delay of the whole package or the need to postpone this item to the next phase (a delay of several years). In order to improve the flexibility and the time to market the concept of the phase 2+ program was developed.

The target was to allow an independant elaboration and introduction of the individual work items. To enable this the modularity of the specifications and the standardization management was improved. A more efficient project management of the individual work item was introduced in 1996/7. In addition a program management was introduced. The production cycles were organised as the production of annual releases. Each release contained an agreed set of work items (typically 20...30). In the meantime the first two releases of the GSM phase 2+ program have been completed: GSM Release '96 and GSM Release '97. GSM Release '98 is elaborated at present. These three releases contain the core of GSM phase 2+. Release '99 will be the first integrated GSM/UMTS release.

The GSM phase 2+ program achieves major goals of the third generation already and evolves therefore GSM from a second generation system to generation 2.5. Key achievements are:

- advanced customized services creation and portability
- data rates up to 100 kbits/s
- higher speech quality
- roaming between satellite, terrestrial and indoor environments

2.3.2 GSM Release '96 completed

GSM release '96 contains 26 work items. Major examples are:

- 14.4 kbit/s data transmission (including n times 14.4 kbit/s)
- SIM lock (including a review under competition law by the European Commission)
- CAMEL phase 1 (service portability and creation based in IN)
- Enhanced Full Rate Codec (EFR)

- HSCSD (High Speed Circuit Switched Data) phase 1
- SIM toolkit
- Support of Optimal Routeing phase 1
- ASCI (Advanced Speech Call Items) phase 1: functions for work groups to be used by the European railways

2.3.3 GSM Release '97 completed

GSM release '97 contains 20 work items. Major examples are:
- CAMEL phase 2: additional IN service creation tools
- GPRS (General Packet Radio Service)
- CCBS (Call Completion to Busy Subscriber)
- ASCI phase 2 (Advanced Speech Call Items)
- SPNP (Support of Private Numbering Plan)

2.3.4 GSM Release '98 under Preparation

It is planned to complete the core specifications of GSM release '98 in February 99. At present 33 work items are under discussion. Major examples are:
- AMR (Adaptive Multirate Codec)
- EDGE (Enhanced Data Rates for GSM Evolution)
- FIGS (Fraud Information Gathering System)
- MNP (Mobile Number Portability)
- MExE (Mobile Application Execution Environment)
- TFO (Tandem Free Operation)

2.3.5 Innovative new Work Items

A surprise is the ever growing flow of innovative new work items, which demonstrates the vitality and evolution potential of the GSM platform. Nearly all will be usable in UMTS. Many are critical for the UMTS success. Examples are:
- AMR (Adaptive Multirate Codec)
- CTS (Cordless Telephony System)
- MExE (Mobile Application Execution Environment)
- SIM toolkit
- JAVA on SIM
- Low volt SIM

- GSM Number Portability
- VHE (Virtual Home Environment)

2.4 ETSI SMG the Workshop for GSM 900 and 1800 Specifications

ETSI the European Telecommunications Standards Institute is a recognised European standardisation Institute. ETSI has more than 500 administrations, network operators, manufacturers etc. as members. There are associate members from non-European countries e.g. Australia, Israel, India, Malaysia, USA. Technical work is done in ETSI Projects (EP) and ETSI Technical Committees (TC).

ETSI TC SMG (Special Mobile Group) is responsible for the GSM specification and standards for use in 900 and 1800 MHz. The ANSI (American National Standards Institute) committee T1P1 is responsible for the GSM 1900 MHz specifications. SMG and T1P1 work closely together. They build their work on a set of common GSM technical specifications.

The ETSI SMG Plenary with 150...200 participants meets three times a year. The technical work is done in the eleven Sub-Technical Committees with 30...150 participants (see diagram below). Most of them created working groups. The total number of participants is about 1000. More than 95% of all technical work is performed by the voluntary contributions.

The TC and STCs are supported by a full-time Project Team. The Project Team provides

- meeting reports
- technical co-ordination
- consistency checking
- updating of specifications

– program and work item management

Table 5. SMG sub-committes.

Sub-Technical Committee	Responsibility
SMG1	Services Aspects
SMG2	Radio Aspects
SMG3	Network Protocols
SMG4	Data Services
SMG6	Operation and Maintenance
SMG7	Mobile Station Testing
SMG8	Base Station Testing
SMG9	SIM Aspects
SMG10	Security Aspects
SMG11	Speech Coding
SMG12	System Architecture

Electronic working is used widely. More information can be found under:
WWW.ETSI.FR/SMG/SMG.htm

3. OUTLOOK TO GSM'S EVOLUTION TOWARDS THIRD GENERATION

The strong growth in mobile networks and Internet points to a large potential of mobile Internet users using mobile multimedia services.

Table 6. Number of users in different networks.

Users in million	1995	2001
Fixed networks	700	1100
Mobile networks	100	350…400
Internet	35	300…500

A part of this potential can be exploited by the GSM Phase 2+ data services. But there remain limitations in the ease of use and the transmission speed. Therefore a UMTS (Universal Mobile Telecommunication Services) concept was developed.

The term UMTS was created in 1986 in the framework of the European Research Program RACE. UMTS should replace as a "Third Generation System" in the long term the "Second Generation Systems" like GSM.

In 1996 the UMTS concept was reviewed and the "old" UMTS replaced by a more market oriented "new" UMTS concept. The following table shows a comparison of the two concepts:

Table 7. Comparison of the two UMTS concepts.

	"Old" UMTS	**"New" UMTS**
Core ideas	Integration of all existing and new services into one new universal network	• focus on innovative services • support GSM services
Partner network	Broadband ISDN	• Intranet • Internet
Introduction	Migration from existing networks	Evolution from GSM and ISDN networks
Global roaming	New development based on INAP	Evolution of GSM roaming based on MAP
Standardization	• FPLMTS as one monolithic standard • work in ITU	• IMT 2000 family concept • work sharing between ITU, ETSI, ARIB/TTC, T1

The UMTS services concept foresees:

- high speed mobile multimedia services
    - wide area mobility:
        - up to 144 kbit/s for fast moving mobiles
        - up to 384 kbit/s for pedestrians
    - low mobility indoor: up to 2 Mbit/s
- advanced addressing mechanisms (e.g. Internet style)
- advanced service creation, customization and portability by the virtual home environment
- seamless roaming between indoor, outdoor and far outdoor environments

In order to ease the introduction dual mode of operation between GSM and UMTS using dual mode terminals are planned.

In order to limit the initial investment the new base station system shall be supported by an evolved GSM core network.

The decision on the basic parameters for the UMTS Terrestrial Radio Access (UTRA) in January 98 leads to an integrated approach:

- paired frequency bands: WCDMA (Wideband Code Division Multiple Access)
- unpaired frequency bands: TD-CDMA (Time Division CDMA)

Table 8. Time table for standardization.

UMTS Standardization Milestone	Target date	Status
UMTS concept and basic parameters	early 1998	completed
UMTS phase 1 specifications	early 2000	
UMTS phase 1 operation possible	in 2002	

The standardisation work follows the time table shown in Table 8.

The main part of standardization work is done by ETSI SMG. ETSI NA6 works on the evolution of an ISDN core network, which can be used together with a UMTS Terrestrial Radio Access Network. Market aspects, regulatory aspects and spectrum aspects are studies by the UMTS Forum. ETSI SMG and the UMTS Forum work together closely.

PART 1

ADVANCES IN SERVICES

Chapter 1

New Speech Related Features in GSM

Steve Aftelak
Motorola GSM Products Division

Key words: GSM, speech, codecs.

Abstract: This section introduces and describes the three most recent speech related features which have been, or are being, introduced into GSM; the Enhanced Full Rate Codec (EFR), Tandem Free Operation (TFO), and the Adaptive Multi-rate Codec (AMR).

1. INTRODUCTION

One might be forgiven for believing that the provision of an acceptable basic speech service in GSM has been accomplished. After all, GSM is now one of the most successful second generation public mobile telecommunications services and most of this success has been built upon implementations of the original RPE-LTP speech codec [1] developed in the '80's. But the service environment continues to evolve. New challenges appear and the system must adapt to provide solutions to them, as we move towards the advent of third generation systems. In this chapter the market and technology drivers which continue to arise are examined, and the solutions developed or under development are described.

1.1 Drivers: Coverage, Capacity, Quality

The cost of launching and maintaining a viable public mobile telecommunications system is immense. Therefore a prime consideration has been, and will continue to be, the provision of service in the most cost-effective manner possible.

Initially (and this is still to some extent true today), the emphasis has been on utilising means to extend and enhance coverage; simply put, how to achieve the desired level of coverage by employing the minimum of costly infrastructure equipment (base stations, base site controllers, switching centres, etc.).

More recently, the emphasis has switched (at least partially) to the question of how to respond to demand for services which has outstripped supply. For many operators, the crisis of insufficient capacity has been particularly acute in city centre areas.

At the same time customers are becoming more discerning; they are beginning to demand better quality in terms of intelligibility, clarity, and absence of the artefacts traditionally associated with radio communications systems. This, allied to the stated aim of many operators to attract customers away from traditional wireline telecommunications service providers, has placed momentum behind moves to improve quality.

The elements which form the core parts of speech service provision play their part to greater or lesser degrees in providing solutions to changing coverage, capacity and quality requirements. Before describing in some detail the developments within GSM, both recently completed and still to be completed which aim to meet emerging needs, it is useful to briefly consider the relationship between core speech technology and these 3 criteria.

1.1.1 Coverage

The enhancement of coverage has not played a major part in the requirements for the more recent speech related features in GSM, but for completeness a short discussion of the relevance of speech and channel coding is provided.

A general rule of thumb for network planning might be the guidelines given in Table 1. This details the nominal average Carrier to Interference ratios (C/I) that should be achieved in various areas.

Table 1. C/I in relation to Cell Area

Nominal C/I (dB)	Area covered
9dB or greater	Well within cell boundary
7dB	Towards edge of cell
4dB	Strictly outside of cell area

In GSM careful design of channel coding is used to achieve acceptable speech quality across a range of static interference environments. In particular, analyses of the sensitivity of speech quality to errors in the coded speech is undertaken, in order to determine which elements in the coded speech require the most error protection. The speech coder output bits are then sorted into subsets according to their sensitivity to error, and more error protection is afforded to the subsets which require it. (For more details see references [1], [2], and [3], which define the error protection classes for the Full Rate (FR) , Half Rate (HR), and Enhanced Full Rate (EFR) codecs respectively.)

It is true to say that enhancements in speech quality can, to some extent, be traded off against coverage gains. As a trivial example, the EFR codec feature (see Section 2.1) was developed primarily in order to improve the overall speech quality experienced by GSM end-users. It is also possible to envisage an operator model which uses the EFR to improve coverage whilst maintaining overall quality at pre-EFR levels. It must be said that this is an inherently difficult task to undertake, and one that is not likely to be widely embraced in conjunction with the EFR codec.

1.1.2 Capacity

The enhancement of capacity via speech codec technology has received some attention in GSM's recent history. The half rate codec was designed to substantially increase capacity by potentially doubling the number of voice calls supported by a single carrier (from eight to sixteen, on the assumption that the carrier is dedicated to carrying traffic channels.)

A disadvantage of speech coding as a mechanism for capacity enhancement is the fact that the gain is capped by the proportion of mobiles operating in half rate mode. This is illustrated by Figure 1, which illustrates the theoretical maximum capacity enhancement achievable with the penetration of half rate capable mobiles as the factor. (This assumes that HR is used at call set-up in preference to FR and that two HR calls are always aggregated into one timeslot where possible, via intra-cell handover. A cell with 22 traffic timeslots is analysed and a 2% blocking probability is assumed.)

The standardisation process for the half rate feature was largely completed by the end of 1994. Nearly three years later, the first commercial implementations of HR are only just beginning to appear. The reasons for the relative lack of success to date are complex, but one point is worthy of note. It is undoubtedly true that the inherent quality afforded by a speech codec operating at a particular rate will always be greater than one operating at around half that rate. Therefore a codec designed for the GSM full rate channel will perform better than one designed for the half rate channel. The speech quality of a half rate solution will always lag behind[1]. A widely held belief has been that users will not accept the poorer speech quality afforded by the use of half rate, but such attitudes may be changing, all be it slowly.

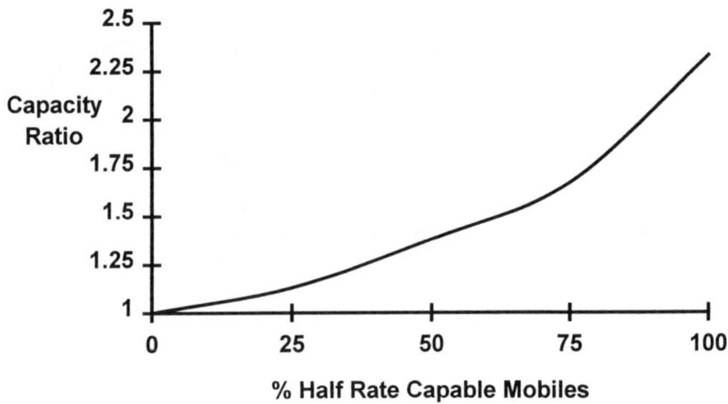

Figure 1 Gain in Capacity compared to FR as a Function of HR Mobile Penetration [2% blocking, 22 available traffic timeslots]

By a somewhat different means, it has been suggested that the Enhanced Full Rate feature may be deployed for capacity enhancement. A side effect apparent with EFR is that its tolerance to interference is somewhat higher than that of the full rate codec system. This is primarily due to the inherent better quality afforded by the speech codec, but is also influenced by the additional channel coding defined for EFR (see section 2.1). As shown in Figure 2, (reproduced from [4]), it is clear that the potential for capacity enhancement becomes small as the interference environment worsens.

1 This argument ignores the inherent quality improvement associated with the lower probability of being blocked at call set-up using HR.

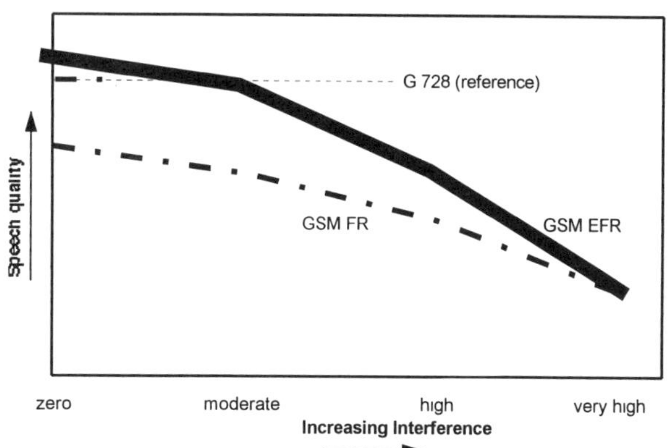

Figure 2 Speech Quality Comparison as a Function of Interference (no background noise). (© ETSI) GSM 06.55 (ETR 305)

1.1.3 Quality

The provision of a good quality speech service is a fundamental requirement for a service such as GSM which relies on voice communication for a very large proportion of its revenue.

Network operators are constantly striving to improve the quality offered to their subscribers. From a technical viewpoint it is to be stressed that this is not just about the search for improved speech coding technology. Core speech codec technology is just one of a large number of factors which govern the quality of the connection offered to the end user. One must also consider (amongst other things) echo control, delay, network optimisation, handover performance, terminal design, edge of cell coverage, and the ability to place calls "wherever, whenever" as factors. (It is interesting to note that many operators would rate echo cancellation performance as the most critical aspect in delivering a quality service. It is also true to say that network optimisation can play a major part in speech quality. In particular the implementation of frequency hopping can via interference averaging improve the average C/I experienced by end-users. This leads to simplified network optimisation whilst allowing tighter frequency re-use, delivering both higher capacity and higher quality.)

From a commercial perspective, whilst accepting that provision of good quality is essential in maintaining satisfied customers, operators often find that the development of a business case to justify investment in a feature solely to enhance speech quality is difficult to develop. Hence the two programmes currently underway to enhance speech quality both also provide other benefits; the Adaptive Multi-rate (AMR) codec feature (see section 2.3) providing capacity and to a lesser extent coverage enhancement, and the Tandem Free Operation (TFO) feature (see section 2.2) providing transmission savings between switches. The recent exception to the rule is the Enhanced Full Rate codec, although the previous sections have addressed its potential use for coverage or capacity enhancement.

2. GSM'S NEW SPEECH FEATURE SET

In this section, the speech related features which are under development for GSM, as well as one feature recently completed, are described. The responsible body overseeing the development of the standards is the technical committee SMG11, *'Speech Aspects'*.

The Enhanced Full Rate (EFR) codec standard was finalised during 1996.

The standard for the so-called Tandem Free Operation (TFO) feature is currently at the detailed design stage, and is due to complete during 1998.

The Adaptive Multi-rate (AMR) codec feature has recently completed a one year feasibility study. This has concluded that the feature will be a viable and useful addition to the portfolio of GSM speech-related features, and a one to two year standardisation phase is underway.

2.1 The Enhanced Full Rate Codec (EFR)

In December 1994, a report was published on the feasibility of introducing a second full rate speech codec into GSM. The report [5] recommended that such a codec should be developed and standardised, and that the codec should be designed primarily for quality. The report suggested that the resulting codec should meet the speech quality requirements set out in Table 2.

Table 2. Original Requirements for the Enhanced Full Rate Codec

Condition	Requirement
Speech Quality - error free	G.728[1]
C/I of 13dB	G.728 (error free)
C/I of 10dB	Significantly better than FR
C/I of 7dB	Significantly better than FR
C/I less than 7dB	Graceful degradation, no annoying effects
End to end delay	No more than FR
Speaker Independence and recognisability	As good as G.728
Tandem EFR-PCM-EFR	G.728-PCM-G.728
Tandem, as above, EP1[2] channel condition	Better than FR-PCM-FR, EP1 channel condition
VAD/DTX[3]	As good as FR
DTMF Transparency	At least as good as FR
Music	Better than FR. No annoying effects

1 ITU-T Recommendation G.728

2 See Section 2.1.3 for a brief description of the error patterns used to determine codec performance over simulated radio channels.

2 VAD/DTX stands for Voice Activity Detection/Discontinuous Transmission, which facilitates the ability to largely avoid transmission on the air interface during silence periods in the source speech

In January 1996 a codec was adopted by ETSI based on Algebraic Code Excited Linear Prediction (ACELP) techniques, a variant of CELP developed by the University of Sherbrooke in Canada. The particular solution was developed jointly by Sherbrooke and Nokia, and had previously been chosen as the preferred codec for operation in GSM systems in North America.

2.1.1 The EFR Speech Codec

The background to CELP codecs, and ACELP in particular, is well documented and is not repeated here. The reader is directed to reference [6], for further information.

Figure 3, reproduced from Reference [3], shows how the speech encoding and decoding functions for the EFR fit together within the complete speech processing function. Note that the GSM specification dealing with each function is marked.

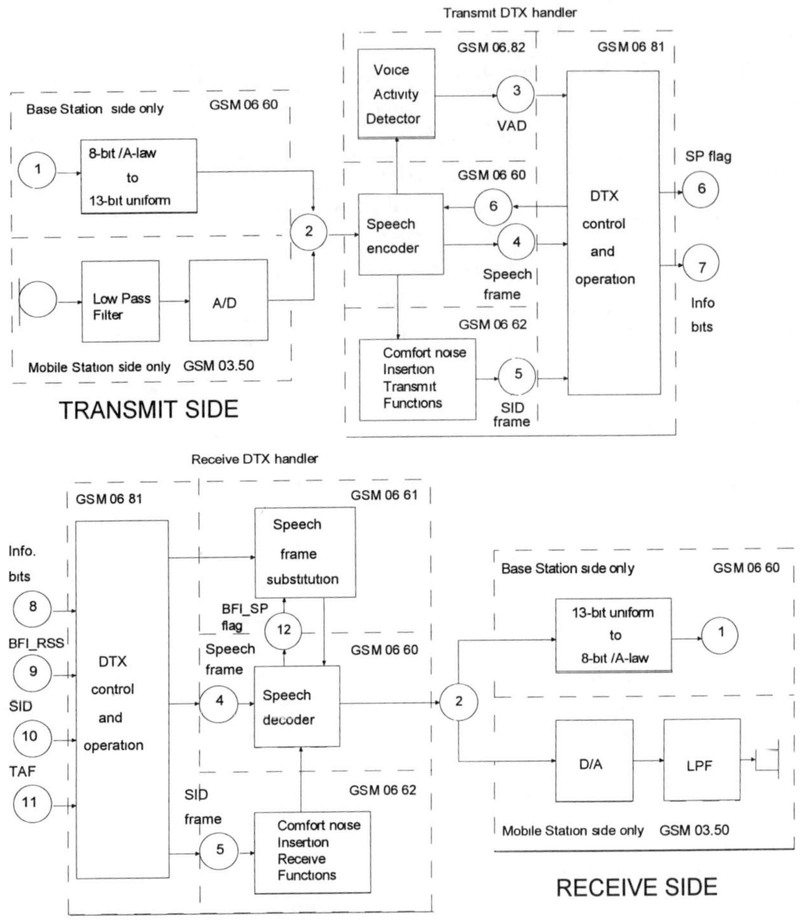

Figure 3 Speech Processing Functions for the EFR Codec (© ETSI) GSM 06.60 (ETS 300 726)

Notes

1 8-bit /A-law PCM (ITU-T G.711), 8000 samples/s
2 13-bit uniform PCM, 8000 samples/s
3 Voice Activity Detector (VAD) flag
4 Encoded speech frame, 50 frames/s, 244 bits/frame
5 SIlence Descriptor (SID) frame
6 SPeech (SP) flag, indicates whether information bits are speech or SID information
7 Information bits delivered to the radio subsystem
8 Information bits received from the radio subsystem

9 Bad Frame Indication (BFI_RSS) flag from Radio Sub System
10 SIlence Descriptor (SID) flag
11 Time Alignment Flag (TAF), marks the position of the SID frame within the Slow Associated Control CHannel (SACCH) multiframe
12 Bad Frame Indication (BFI_SP) flag from speech decoder

The VAD/DTX and Comfort Noise Insertion functions are not detailed here, since they are broadly based on previously used techniques. See the relevant GSM specifications for more details.

The speech encoder operates on speech frames which are 20ms in length, which is equivalent to 160 samples at the 8000 samples/s sampling rate.

At the core of the speech encoder are two all pole synthesis filters which model speech production as an autoregressive (AR) process, along with a model of the excitation process. A 10th order linear prediction filter is used as defined in Equation 1, where a_i, i=1,...m, are the quantized linear prediction parameters.

$$H(z) = (1 + \sum_{i=1}^{10} a_i z^{-i})^{-1} \tag{1}$$

The second filter is the pitch or long-term synthesis filter as defined in Equation 2.

$$B(z)^{-1} = (1 - g_p z^{-\tau})^{-1} \tag{2}$$

where τ is the pitch delay (or lag) and g_p is the pitch gain. The implementation of the latter filter is via the adaptive codebook approach. Firstly a search is undertaken to find the optimal time lag τ. This is initiated with an open-loop pitch search. This simplifies the task by confining the subsequent closed-loop pitch search to a small number of points around the open-loop estimated value. The appropriately time-shifted (using τ) and amplitude-scaled (using g_p) block of previous excitation samples is taken to be the current excitation component associated with the pitch, thereby eliminating the pitch synthesis filter.

The excitation signal at the input to the short term linear prediction synthesis filter is formed by adding the excitation vector produced by the adaptive codebook approach to an appropriate vector chosen from the fixed codebook. The choice of codebook vector is realised by undertaking a search with the aim of minimising the error between the original and synthesised speech, where the error is measured using a perceptually weighted distortion measure. This process uses the weighting filter of Equation 3.

$$W(z) = A(z/0.9)/A(z/0.6) \quad (3)$$

Here $A(z)$ is the unquantised linear prediction filter, whereas the short term synthesis filter $H(z)$ uses quantized filter coefficients. The ACELP approach lends itself to an efficient codebook search procedure utilising nested loops [6]. Each code vector consists of 10 non-zero pulses. Each pulse can take on one of eight positions and can take on the value +1 or -1. By grouping pulses in pairs, each pair can be encoded using 6 bits for position information (i.e. 3 bits per pulse) and only 1 bit to encode the sign. The sign bit defines the sign of the first bit of the pair. The sign of the second bit of the pair is defined by its position relative to the first bit. As a result, only 35 bits are required to define a codebook vector.

Linear prediction analysis is undertaken twice per speech frame. The resulting linear prediction coefficients are converted to Line Spectrum Pairs (LSPs) and are jointly quantised such that they are represented by 38 bits. The speech frame is split into four subframes of 5ms each in length. The calculated linear prediction coefficients are directly associated with the second and fourth subframes. Interpolation is used to generate the linear prediction coefficients for the first and third subframes. A new set of adaptive and fixed codebook parameters is calculated for each subframe, but open loop pitch lag estimates are generated only twice per speech frame (i.e. every 10ms). The pitch delay is coded with 9 bits for the first and third subframes, but is coded relative to the previous values for the second and fourth subframes, thereby reducing the required bit allocation to 6 bits per delay value. The gains for the adaptive and fixed codebooks are quantised to yield 4 and 5 bit quantities respectively.

Table 3 gives the bit allocation of the speech codec. It generates 244 bits per speech frame, to be forwarded to the channel coder.

Table 3. Allocation of Bits to EFR Encoder Output Parameters

Parameter	**1st & 3rd subframes**	**2nd & 4th subframes**	**total per frame**
Jointly Quantised LSPs			38
Pitch delay	9	6	30
Pitch gain	4	4	16
Algebraic code vector	35	35	140
Algebraic Codebook gain	5	5	20
Total			244

At the decoder, the received parameters are used to synthesise reconstructed speech. The LSPs are decoded to yield the two sets of linear prediction coefficients. Interpolation is undertaken to yield four sets of coefficients; one set per subframe. For each subframe, the excitation is constructed by adding the fixed and algebraic codebook vectors scaled by their respective gains, and the speech is reconstructed by filtering the resultant excitation signal through the appropriate synthesis filter. The resultant signal is filtered by an adaptive post-filter. It is well known that this can reduce the rough quality of the speech at the output of CELP decoders [6]. In this case the post-filtering is composed of two parts; a formant post-filter, and a tilt compensation filter. The former is based on the linear prediction filter. The formant post-filter is given by equation 4.

$$H_f(z) = \frac{A(z/\gamma_n)}{A(z/\gamma_d)} \tag{4}$$

$A(z)$ is the received quantised (and interpolated) linear prediction inverse filter. The second filter compensates for the spectral tilt introduced by the formant postfilter and is given in Equation 5.

$$H_t(z) = (1 - \mu z^{-1}) \tag{5}$$

where $\mu = \gamma_t k_1'$ is a tilt factor, with k_1' being the first reflection coefficient calculated on the truncated impulse response of the formant post-filter.

2.1.2 Additional Channel Coding for the EFR Codec

Channel coding for the EFR feature is heavily based on that used for the original Full Rate codec. The same forward error correction scheme is used, but this is preceded by a layer of extra protection, called preliminary channel coding [7]. This affords a measure of extra error detection capability compared to the FR codec, and arises from the fact that the EFR speech codec rate is lower than that for the full rate speech codec (12.2kbps, rather than the 13kbps). Therefore extra channel coding, to the tune of 800bps (or equivalently 16 bits per speech frame), can be introduced. This preliminary coding is described in some detail. The follow-on error protection coding is described less fully since it is the same as that used for the full rate codec; for full information see [7] or a standard text such as [6].

The 244 data bits out of the speech codec are split into three classes; Class 1a (50 bits), 1b (124 bits), and 2 (70 bits), with Class 1a deemed to be the most important in terms of the quality of the speech. The output of the preliminary coder comprises of 260 bits. The preliminary coding operates as depicted in Figure 4.

Figure 4 Preliminary Channel Coding for the EFR Codec

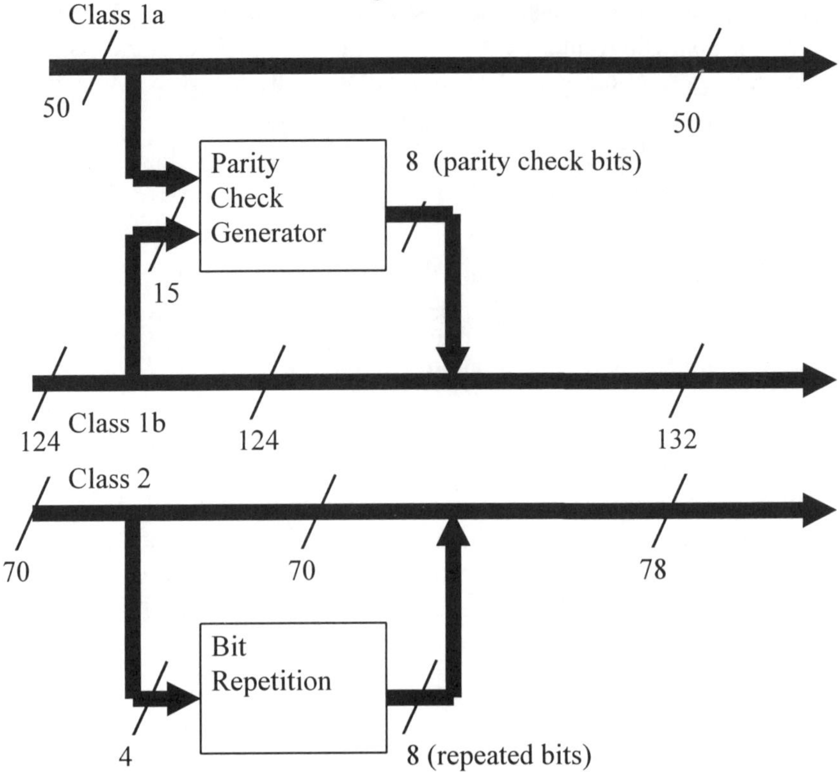

Eight parity check bits are generated from the Class 1a bits and the 15 most important Class 1b bits, using the cyclic generator polynomial: $g(D) = D^8 + D^4 + D^3 + D^2 + 1$. These bits are included in Class 1b to give a total of 132 bits.

Additionally, the 4 bits out of the speech coder which are deemed to be the most important bits of Class 2 are protected in a very simple manner by repeating each bit twice so that, for example, a simple majority logic decoding scheme can be used at the receiver to decode them. These bits are included in Class 2 to give a total of 78 bits.

The standard GSM full rate channel coding then comprises of the generation of 3 parity check bits using the Class 1a bits, followed by half rate convolutional encoding of the resulting 50 Class 1a, 132 Class 1b, 3 parity check, and 4 tail bits, to give a grand total of 456 bits (378 bits at the output of the convolutional encoder together with the 78 Class 2 bits).

2.1.3 Performance of the Enhanced Full Rate Codec

This section concentrates on the speech quality aspects of EFR codec performance. (Use of the EFR feature for coverage or capacity enhancement was discussed briefly in section 1.)

Formal listening tests were undertaken as part of the development of the EFR codec standard. A number of tests were defined to test performance under a range of interference scenarios, in background noise conditions, and for tandem connections (where two codecs are effectively cascaded in a mobile to mobile call).

The interference scenarios are categorised in terms of error patterns which simulate particular conditions. Pattern EP0 (no errors) simulates conditions near the centre of the cell. EP1 is meant to simulate conditions found in the bulk of the cell (C/I of 10dB) whereas EP2 simulates conditions near to the edge of a cell. Finally, EP3 corresponds to locations which are, strictly speaking, outside of the serving cell. In reality, the dynamic nature of the radio channel means that condition similar to those simulated by EP2, and even EP3, can be encountered well within a cell and in general a call may experience a range of interference environments during its course. Figure 5 is a comparison of codec performance across this set of conditions, from tests undertaken by Motorola. The measure of quality used is the Mean Opinion Score (MOS), which is widely used in subjective tests. Briefly, in this type of test listeners are asked to rate speech according to the scale of Table 4. For more details see Reference [8].

Table 4. Definition of the Mean Opinion Score Scale

MOS value	Meaning
5	Excellent
4	Good
3	Fair
2	Poor
1	Bad

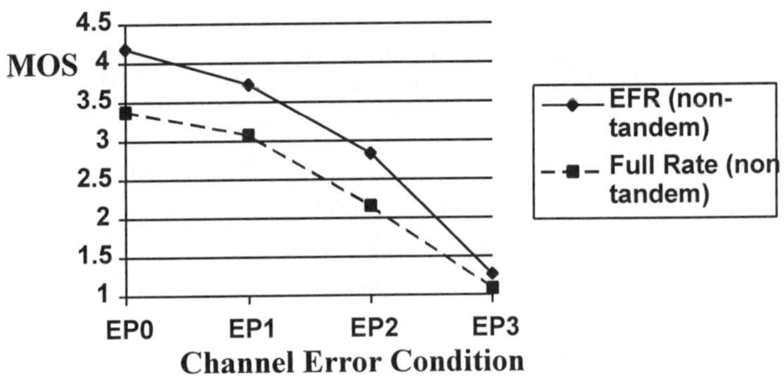

Figure 5 Comparison of EFR and FR Codec Quality with Channel Errors

Figure 5 shows that the quality of EFR codec in conditions of light interference is indeed superior to that of the FR codec, but that this advantage is eroded as the interference becomes stronger. This is inherently due to the fact that the EFR feature largely retains the error protection system of the FR codec.

Examples of the performance of the EFR codec in conditions of background noise are given in Table 5. These are taken from subjective tests undertaken as part of the development of the standard [4]. It should be noted that independent experiments were undertaken for each type of noise, and that these experiments were also independent of those undertaken to yield Figure 5. The decorrelation between the absolute MOS scores recorded in the experiments is due to this. (In fact, comparisons on the basis of MOS should never be undertaken between independent experiments.)

*Table 5.*Comparison of EFR and FR codec Performance in Background Noise

	MOS in given Background Noise	
	In Vehicle (10dB SNR)	Music (20dB SNR)
EFR	4.4	4.3
FR	4.2	3.4
G728	4.5	4.5

The tests included the ITU-T G.728 codec, which is considered to be a high quality reference. It can be seen that the EFR codec performance is very similar to that of G.728. (In fact, the results are statistically inseparable.) Equally, it is clear that the EFR performs at least as well as the FR codec for in-vehicle noise, and performs considerably better with music in the background.

Table 6 gives examples of EFR performance when self-tandemed (i.e. simulating a mobile to mobile call where both handsets employ the EFR codec.) The tests were undertaken over the EP1 channel.

Table 6. Comparison of EFR and FR codecs when Tandemed

Codec	MOS (EP1)
EFR	2.6
FR	2.3

Again, it is clear that the EFR codec has better performance than the FR codec, under the condition tested [4].

2.2 Tandem Free Operation

It has long been known that the cascading of two or more speech codecs (i.e. encode - decode - encode -decode ...) can lead to serious degradation in the quality of the reconstructed speech, and GSM is one example of this phenomenon (for mobile to mobile calls). Therefore, during 1996, a work item was started under the auspices of SMG11, in order to devise a solution to this problem.

This work is still underway, with the aim of finalising the standard during 1998. Therefore this description is largely restricted to the principles of operation, rather than a complete definition of the solution.

The principle aim of Tandem Free Operation (TFO) is simple; ensuring that only one speech codec (i.e. speech encode - decode operation) operates for mobile to mobile calls. This is depicted graphically in Figure 2.1.6, where it is assumed that the reader has a rudimentary knowledge of GSM system architecture. During the operation of TFO it is the mobile stations alone which encode and decode speech. Speech is transported in encoded form along the remainder of the path between the two end users. Thus, in place of the transport of PCM-encoded speech between the transcoder equipment, across the MSCs, GSM-encoded speech (i.e. FR, HR, or EFR encoded speech) is transported.

The network entity logically ascribed the responsibility of managing and implementing TFO operation is the transcoder, named the Transcoder and Rate Adaption Unit (TRAU) in GSM. This equipment, amongst other things, performs speech encoding and decoding. It is worth noting that in Figure 6 this unit is placed at the MSC site, although it is permissible for it to be placed nearer the air interface within the Base Site Subsystem.

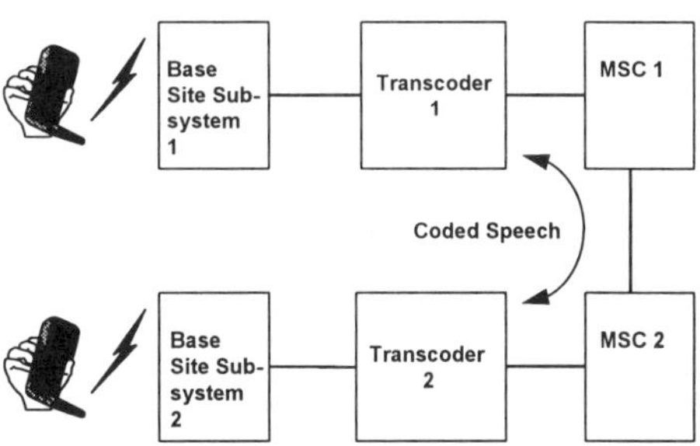

Figure 6 The TFO Concept

Table 7 lists the more important requirements for tandem free operation (TFO) as defined by SMG11 [9].

Table 7 indicates that great emphasis is placed on reducing or eliminating any speech quality degradations due to TFO operation, and for good reasons. Firstly, the principle aim of TFO is the improvement of speech quality, so any effects, all be it short term, which reduce speech quality to a noticeable extent are to be avoided! Secondly, the mechanisms defined to set-up TFO will also operate during the beginning of GSM calls where one party is in the fixed network (there being no currently defined mechanism for the system to know this in advance). Since at present and for the foreseeable future, most calls using the GSM network do involve one party in the fixed network, it would be unacceptable to lower the quality of such calls through the operation of TFO, even if the degradation is of very short duration. Additionally it indicates that importance is associated with the need to transit (or revert) quickly to normal dual-codec tandem operation when needed. This is discussed in more detail in the following sections.

In the following sections, the operation of TFO is described firstly in respect of the mechanisms necessary to establish it. The continuous operation of TFO is then discussed, together with the mechanisms used to detect the need to revert back to normal (tandem) operation. These descriptions are of necessity brief. A more detailed description may be found in Reference [10]. A brief analysis of the speech quality benefits to be expected is provided. Finally, a secondary, yet potentially important, advantage of TFO is discussed.

Table 7. Central Requirements for TFO

Area of Impact	Requirement
Applicability	All mobile to mobile calls using the same speech codec
Effect on Mobile Station	No modifications required
Dependence on location	Both inter- and intra-network calls possible (operator choice)
Subscription/billing	No impact
Time to set up TFO following call set-up	2s (objective), 7s (maximum) No perceptible speech degradation during this phase.
Time to re-enable TFO after an interruption (e.g. a handover)	7s (objective). No perceptible speech degradation during this phase.
Time to return to normal (i.e. tandem) operation when required	160ms.
Transitions to/from TFO mode of operation	No annoying artefacts
Continuous employment of TFO	No degradation
Transition to multiparty call	Revert to normal (tandem) operation
Supplementary services in general	No interference with service provision
Lawful Intercept	No degradation in speech quality when call is intercepted

2.2.1 TFO Establishment

At the start of this phase, TFO-compliant TRAU equipment will emit messages on the A interface towards the MSC to which it is connected, announcing that it is TFO capable. A TFO capable TRAU will begin to emit this message as soon as (and at the same time as) it begins to send decoded speech towards the MSC.

The physical means of transport is not yet finalised, but it seems very likely that it will use a form of bit stealing, similar to that used to carry signalling on T1 digital trunk lines in North America [11]. Briefly, this involves the regular appropriation of the one LSB of a PCM octet in every

N such PCM octets, in order to build up the message (where N is likely to be set to 16).

The exact content of this 'TFO Request' (TFO_REQ) message is also to be fixed, but it may contain little more than a basic header which effectively announces that the emitting TRAU is TFO-capable. By reducing the content, and therefore length of the message, the effect of this embedded message on the quality of the transported speech is considerably reduced.

A TFO capable TRAU, on receipt of a TFO_REQ message, will respond with a 'TFO Acknowledge' (TFO_ACK) message which imparts considerably more information as well as acknowledging receipt of the TFO_REQ. The information is contained within extension blocks which inherently indicate whether further extensions follow by virtue of a 'more' field. The TFO_ACK message includes an extension block defining the speech codec type in use, and optionally a block defining the speech codecs which may be used. The latter block is used to (optionally) facilitate negotiation of a common codec to be used for the end to end connection. Such negotiation will require signalling support within each base station sub-system participating in the call, but it is envisaged that the latter will not be standardised. The alternative to negotiation upon codec mis-match is to abandon the establishment of TFO. Once both TRAU units have exchanged TFO_ACK messages which record that both are using the same speech codec, TFO Operation state is entered where encoded speech is transported on the A interface. This procedure is shown diagramatically in Figure 8.

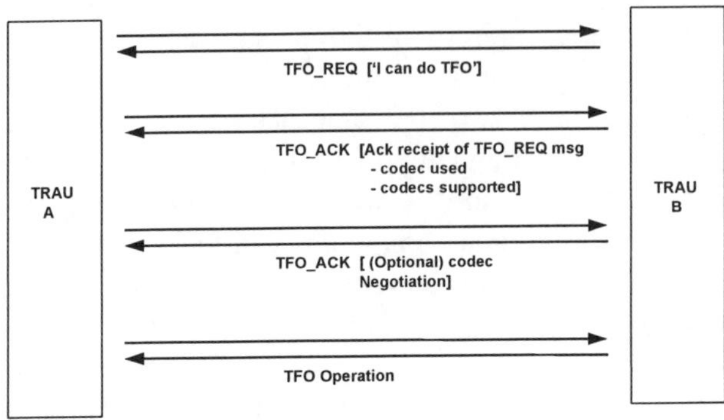

Figure 8 TFO Establishment

Additionally, the TFO establishment mechanisms include means of controlling the operation of in-path equipment which may, by virtue of their operation, obstruct the successful set-up of TFO. Such equipment includes echo control and speech compression devices which render the path between the TRAU units non-transparent. TFO_ACK messages will contain a 'command' field dedicated to controlling compliant in-path equipment. Examples of the genre of commands envisaged include 'pass this complete message' and 'pass the lower 2 bits of all subsequent PCM words transparently until further notice'. The in-path equipment will need to detect the defined TFO header field in order to subsequently detect and decode commands.

2.2.2 TFO Operation

GSM defines a framed transmission format for coded speech on the interface between the base station (BTS) and TRAU unit on the so-called Abis interface, where such frames are known as TRAU frames. (As an example see Reference [12]). This format has been re-used in TFO, with some amendments [10], for the transport of coded speech between TRAU units via the A interface. Such frames are designated as TFO speech frames. TRAU units will search for TFO speech frame synchronisation on the A interface in a similar manner to that currently employed on the Abis interface. These frames are transported on the A interface by replacing the LSBs of PCM samples with the TFO speech frames (one LSB for HR, two LSBs for FR and EFR).

Once a TRAU unit has both received and transmitted an appropriate TFO_ACK message (see Section 2.2.1), it will start to transmit appropriately coded speech in TFO speech frames. It will also search for incoming TFO speech frames whilst continuing to encode PCM for transmission on the downlink leg of the transmission path. When TFO speech frames are received, they are forwarded (with appropriate amendments) on the downlink leg as TRAU frames in place of encoded speech derived from the received PCM.

The need to revert to normal (i.e. non-TFO) operation is achieved by the detection of the loss of TFO speech frame synchronisation in the downlink TRAU receiver. In order to facilitate a fast transition to normal (non-TFO) operation, those bits of each PCM sample which do not carry the TFO speech frame information continue to carry the PCM sample bit values (i.e. the upper 6 or 7 bit values). Thus, as soon as the receiver detects a need to revert back to normal operation, it can immediately use the received samples as input to the speech encoder in place of forwarding the received TFO speech frames on the downlink. It is envisaged that the speech encoder will operate continuously during TFO operation to allow this.

2.2.3 TFO Performance Evaluation

SMG11 undertook some limited subjective tests, based on simulations of TFO operation, which show the level of improvement expected. This is provided in Table 8, which gives the expected improvement in MOS score (strctly speaking DMOS score, see Reference [8]) for TFO compared with

Table 8. Expected Improvement (MOS) due to TFO

Codec	Channel Condition		
	EP0	EP1	EP2
HR	0.9	0.7	0.4
FR	0.5	-	-
EFR	0.3	0.5	0.2

self-tandemed operation (e.g. the EFR codec cascaded with the EFR codec). It is clear from Table 8 that the potential gains are significant.

2.2.4 Transmission Savings using TFO

Since the transport of TFO speech frames on the A interface utilises only one eighth or one quarter of the capacity of the PCM channel, it is useful to consider the potential use of a submultiplexing scheme where four (EFR or FR) or eight (HR) speech channels can be accommodated on one 64kbps link. By this means significant savings in transmission resource can be gained. This is indeed so, and the TFO standard will include sufficient mechanisms to accommodate this. It is true to say that such submultiplexing is being implemented today in GSM for inter- MSC links, but this utilises non-GSM speech encoding standards such as G.728, and therefore may add to degradation by inserting extra tandem stages. By integrating this multiplexing into TFO, the same is achieved without extra tandem stages, and indeed for mobile to mobile calls with only one speech codec in operation end to end.

The principle is quite simple. Each end of an inter-MSC link would incorporate a device which is similar to a TFO-capable TRAU, in that it contains the GSM-standardised speech codecs, and it is capable of interpreting and (possibly) generating TFO messages. The default operation of the devices might be to encode and decode using the EFR codec, so that submultiplexing is permanent on the inter-MSC link. Mobile to fixed phone calls using the inter-MSC link would, by negotiation if necessary, utilise the same speech codec in the GSM TRAU device and the inter-MSC devices, with only the inter-MSC device at the MSC connecting to the fixed network performing transcoding; i.e. the GSM TRAU and intermediate inter-MSC device simply relay the coded speech. In mobile to mobile calls, both the inter-MSC devices and the GSM TRAUs would act as relays for the coded speech.

2.3 The Adaptive Multirate Codec (AMR)

In an important sense, the title of this feature is misleading in that it may lead the reader to assume that this is yet another speech codec or set of speech codecs for GSM. The reality is very different, and the introduction of this feature will affect, to greater or lesser degrees, all the GSM network components which directly carry the speech data.

It will be necessary for the AMR solution to be capable of operating both in the half rate GSM channel where the aggregate bit rate for speech and channel coding is 11.4kbps ([2], [7]), and in the full rate GSM channel where the aggregate bit rate is 22.8kbps ([1], [7]). Half and full rate operation are termed channel modes. Within each channel mode, it is envisaged that the split of the available data rate between speech and channel coding will be variable. Each defined combination of speech and channel coding is called a codec mode. (Currently there is no clear indication as to the likely set of codec modes.)

It is envisaged that a new set of speech and channel codecs will be required in order to define the set of codec modes.. The intention is two fold:

1. To accrue capacity advantage in low interference conditions via the predominant use of the half rate channel mode.
2. To accrue quality advantages in high interference conditions by the use of the full rate channel mode with a high ratio of channel coding to speech coding.

The choice of operating state (half rate or full rate, plus associated codec mode) may be determined by a number of factors (e.g. congestion, time of day) , but it is envisaged that it will be governed primarily by signal quality measurements made by the system in real (or near real) time. In band signalling will be used to communicate the required measurements (or measurement results) to the system component(s) controlling the mode of operation.

The system entity or entities (to be defined) which control the choice of rate then need to communicate the choice to the system components which need to react to the choice (e.g. transcoders, channel coders). Changes of channel mode (e.g. half rate to full rate) will utilise already defined handover procedures using out of band signalling. On the other hand codec mode changes within a fixed channel mode will be communicated in band.

Studies undertaken thus far have indicated that channel mode changes, implemented via intra-cell handovers, are likely to occur a few times per minute [13]. Such changes will be symmetric in the sense that both directions of speech transport always use the same channel mode. On the other hand it is envisaged that changes of codec mode will occur much more frequently (up to a few times per second), and that asymmetry of choice is allowed, so that different codec modes can be chosen for uplink and downlink depending on the prevailing channel conditions in the two directions.

Table 9 defines the more important requirements provisionally defined for the AMR codec [13].

Table 9. Important Requirements for the AMR Feature

Channel Mode	Condition	Requirement
Full Rate	EP0	EFR at EP0
	EP3	EFR at EP1
	Background Noise (EP0)	EFR
Half Rate	EP0	FR at EP0
	EP1	FR at EP1
	EP3	FR at EP3
	Background Noise (EP0)	FR at EP0

Figure 8 illustrates the potential performance benefits (theoretical) which the AMR feature may offer, both in terms of capacity and speech quality [13]. It provides curves of speech quality [Q values in dB, which are derived from MOS results as described in [8]) against capacity advantage, for a possible AMR codec and for a combination of the use of the EFR and HR codecs. It also gives an indication of the current quality offered by the FR codec. It is to be stressed that the performance comparison for the solution eventually chosen may vary considerably from this. (The data of Figure 8 was generated using a radio system simulator, assuming 100% penetration of AMR or EFR/HR codecs, ideal frequency hopping and power control, in a system where each cell uses four carriers.)

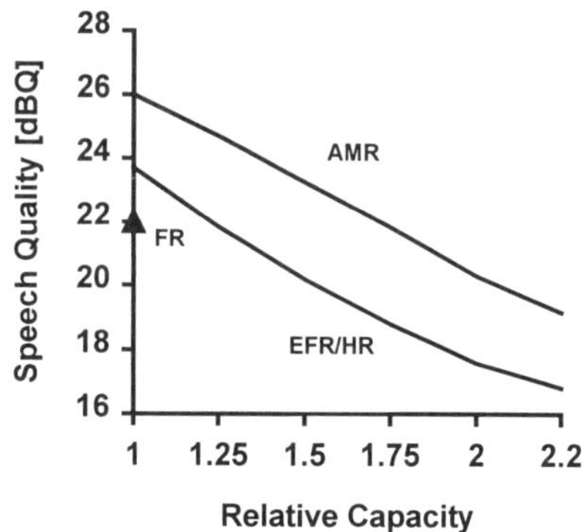

Figure 8 Hypothetical performance Comparisons for the AMR Codec (© ETSI) "Adaptive Multi-rate (AMR) Study Phase Report", SMG11, October 1997

From Figure 8 it is clear that the AMR promises significant improvements. For example, an operator requiring capacity improvements can achieve close to a 75% improvement in capacity without degrading average voice quality in comparison with the FR codec.

The price to pay for the advantages the AMR has to offer is increased complexity in equipments along the speech carrying path in the GSM network. An AMR compliant mobile station will require new speech and channel codecs, enhanced measurement and channel quality reporting functions, and extensions to call set-up and handover functionality. The infrastructure will need upgrades to the BTS (channel codec and TRAU frame format), transcoder (speech codec and TRAU frame format), BSC (radio resource changes for packing half rate calls optimally into the available channels on the air interface, including enhanced handover and channel assignment algorithms). There may also be some changes in the MSC.

A feasibility study into the concept of the AMR codec has concluded favourably. A standardisation process has begun, with a view to publishing the core specifications (to facilitate the start of mobile terminal development) at the end of 1998. The remaining specifications will be completed during 1999. Currently the thinking is that the AMR codec may indeed become the basis for basic voice service provision in UMTS (the European third generation cellular system).

3. CONCLUSION

Experience with GSM has shown that there is a need for new speech related features to play their part in the continuing quest to improve coverage, capacity and quality of digital cellular systems. This chapter has examined this is terms of a recently completed standard (the EFR codec), a standard which is nearing completion (Tandem Free Operation), and a possible future standard which may well form a bridge to the European third generation system, UMTS.

ACKNOWLEDGEMENTS

Many thanks to Dr. Andrew Aftelak, Dr. John Gibbs, Dr. Rene Jepson, Mr Andy Wilton, and Mr Steve Doyle for their review comments.

The present document includes parts of the ETSI technical standards ETR 305 and ETS 300 726, which are property of ETSI.

The original version of the ETSI technical standards can be obtained from the Publication. These standards are available from the ETSI publication Office of ETSI: Tel. +33 (0) 92 94 42 41, Fax: + 33 (0) 4 93 95 81 33.

Requests for authorization to make other use of the documents or otherwise distribute or modify them need to be addressed to the ETSI Secretariat, Fax: +33 493 65 47 16.

REFERENCES

1 GSM 06.10 (ETS 300 580 2): "Digital cellular telecommunications system; Full rate speech transcoding".

2 GSM 06.20 (ETS 300 969): "Digital cellular telecommunications system; Half rate speech; Half rate speech transcoding".

3 GSM 06.60 (ETS 300 726): "Digital cellular telecommunications system; Enhanced Full Rate (EFR) speech transcoding".

4 GSM 06.55 (ETR 305): "Performance Characterisation of the GSM Enhanced Full Rate (EFR) speech codec".

5 ETSI SMG1 Committee, “Enhanced full-rate speech Study Phase Report”, 19th December 1994

6 R Steele (Ed.), "Mobile Radio Communications", Pentech Press, 1992.

7 GSM 05.03 (ETS 300 575): "Digital cellular telecommunication system (Phase 2); Channel coding".

8 ITU-T Recommendation P.830, " Subjective Performance Assessment of Telephone-band and Wideband Digital Codecs ".

9 GSM 02.53: "Digital cellular telecommunication system (Phase 2+); Tandem Free Operation (TFO); Service Description; Stage 1"

10 GSM 04.53: "Digital cellular telecommunication system (Phase 2+); Inband Tandem Free Operation (TFO) of Speech Codecs; Service Description; Stage 3"

11 ITU-T Recommendation G.704, "Synchronous Frame Structures used at Primary and Secondary Hierarchical Levels".

12 GSM 08.60 (ETR 300 737): "Digital cellular telecommunications system (Phase 2+); Inband control of remote transcoders and rate adaptors for Enhanced Full Rate (EFR) and full rate traffic channels".

13 Chairman, SMG11, "Adaptive Multi-Rate (AMR) Study Phase Report", October 1997.

Chapter 2

Advanced Speech Call Items

William Webb
Netcom Consultants

Key words: GSM, group calls, broadcast calls, priority, pre-emption, railways.

Abstract: This section describes the advanced speech call features, which are group and broadcast calls and priority and pre-emption. These services were standardised at the request of the international railway body but are available to all users.

1. INTRODUCTION

The impetus for the Advanced Speech Call Items (ASCI) came from the international railway body, the Union Internationale des Chemins de Fer (UIC). The UIC is the body responsible for harmonising railway standards throughout the world, although its activities tend to be concentrated within Europe. At present, most railways in Europe have incompatible radio systems such that if trains travel from one country to another they typically need to be equipped with more than one radio. Current radios are analogue systems designed in the 1970s and the UIC saw the opportunity to upgrade these to digital radios at the same time as harmonising the use of radio across Europe. The UIC spent some time deciding whether the new radio system should be based on GSM or on an emerging pan-European standard for private mobile radio (PMR) known as the Trans-European Trunked Radio (TETRA). The reasons why the UIC selected GSM are many and varied and are discussed in more detail in [Webb]. Of importance to the evolution of GSM into UMTS - a system able to support a wide range of different users - is that the railways considered that with the ASCI features GSM would provide sufficient functionality for PMR users, a group who historically used specialised radio systems.

The ASCI features are:

the voice broadcast service (VBS) which allows a single mobile to talk to a group of mobiles;

the voice group call service (VGCS) which allows a group of mobiles to talk amongst themselves;

enhanced multi-level priority and pre-emption (EMLPP) service which allows callers with urgent calls to pre-empt those with less urgent calls.

The work items were introduced in the Phase 2+ list in October 1993. The first SMG3 ad-hoc meeting on ASCI was held in January 1994, and meetings have been held regularly since, with the specification work mostly being concluded in January 1997. The requirements and outline descriptions defining the architectural issues were approved by SMG in July 1995 and the major components of the stage 3 descriptions were approved by SMG in April 1996.

The basic requirements and function of each of these services is now described in detail, followed by a discussion as to the key architectural decisions taken during the design of the services. Firstly, a few terms, used in the description are introduced. An understanding of these will make the subsequent explanation simpler to follow.

- **Anchor MSC**. For each group or broadcast call, a particular MSC is designated to handle that call. This MSC is known as the anchor MSC.
- **Despatchers**. These are particular users, often connected via the PSTN, who have high levels of privileges to interrupt and terminate calls.
- **GCR**. The group call register. This is a database containing information about the coverage area of the group call and other necessary factors.

•**Group identity**. This is a phone number, with the structure of an IMSI which is dialled to indicate the group to which the group or broadcast call is intended.

•**Notification**. This is the group or broadcast call equivalent of paging. These are messages notifying mobiles of a group or broadcast call. Unlike paging, the mobiles are not expected to respond to these messages.

•**Service areas**. These are the areas within which the group or broadcast call is to be broadcast.

2. THE VOICE BROADCAST SERVICE

2.1 Overview of the service

The aim of the voice broadcast service (VBS) is to provide a facility whereby either a mobile user or a fixed user (i.e. one connected via the PSTN) can broadcast a voice message to a number of listeners. The listeners are:

- all the mobiles who are members of the broadcast group, who are within a pre-defined coverage area and who are able to accept the call;
- all the pre-defined fixed users.

Essentially, a particular number dialled by the originator of the call corresponds to a particular group and a particular area, described in terms of the cells which cover that area. The call is broadcast within all cells in that area, and any mobiles who are members of the group are entitled to listen. In addition, a number of users not in the GSM system but instead using fixed phones connected via the PSTN can also be connected into the call. The telephone numbers of these users are stored along with the cells in which coverage is required for a particular broadcast call.

A diagrammatic representation of a broadcast call and the architecture established to provide the call are shown in Figure 1, where a mobile sends a group call request to the GSM network. This consults the group call register to determine the service area and the fixed users required and then requests base stations to establish the call and start the notification process.

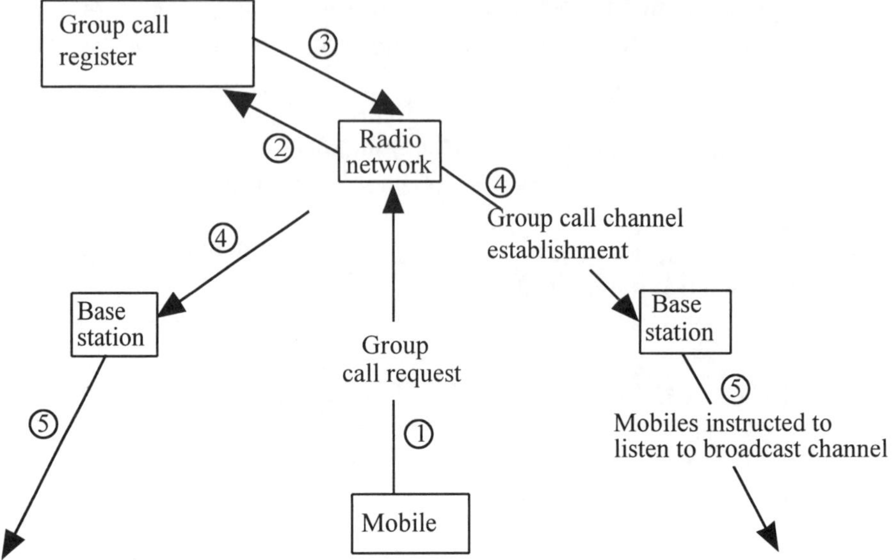

Figure 1 - Schematic showing broadcast or group call establishment

The originator can send his message and then terminate the call. The originator is not aware of who is listening to the call but can request acknowledgements to be sent after the call is completed.

If a voice broadcast call is originated by a mobile, the cell from which the call was originated and the number called are passed to a new device, called the group call register (GCR). The GCR will be explained in more detail later but is essentially a database, typically located within the MSC, which contains stored information about group and broadcast call parameters. The GCR has an entry for each number called, which is called a group identity, and for each group identity it has an entry for each of the cells within which calls for that group identity might be started. When presented with the dialled identity and the originating cell, the GCR returns to the MSC a list of cells within which the call is to be broadcast as well as a number of other parameters which are discussed later. The MSC instructs the required BSCs to establish a single broadcast channel in each affected cell and to commence notification. Mobiles receiving the notification move to the broadcast channel and listen to the downlink. The originator remains on the dedicated channel he used to originate the call and terminates the call when he has finished talking.

2.2 Particular features

Handover

This is straightforward. Since the broadcast call originator is on a dedicated channel, he is handed over in the same way as a mobile in a point-to-point call. The listening subscribers hand themselves over (more correctly they perform cell re-selection) using the same algorithms as idle mode cell re-selection but using a new BA list, which is the list containing details of the surrounding cells. There was much discussion as to whether the BA list should contain only those cells (if any) also carrying the group call but it was eventually decided that this might cause the mobile to move to an inappropriate cell where, if they then requested the uplink, they would cause excessive interference. For this reason, the BA list is mostly the same as it would be for a mobile performing idle mode cell re-selection. However, there is the option to add information as to where the notification channel can be found in surrounding cells, speeding the cell re-selection process and thus reducing potential breaks in speech. Such an enhanced BA list would then contain:

- the neighbour cells frequencies and BSICs, the latter ensuring that the correct cell is chosen
- C1 & C2 parameters which improve the performance of cell re-selection.
- the NCH position to improve handover performance.

Late entry into broadcast calls

Since a broadcast call covers a particular area, described in terms of cells, it is possible that users who are entitled to hear the call will move into the area during an on-going call. This is entitled "late entry". The requirements are that these users should be able to listen to the call. In order to allow this, notification messages are periodically sent on the notification channels throughout the duration of the call. In this way, late entrants are able to find out about the call and move to the appropriate channel.

Encryption

A simple encryption process is available whereby a number of cipher keys per group are provided. These are stored on the SIM so when a mobile enters a broadcast call it retrieves from the SIM the cipher key indicated in the notification message and uses it directly. All cells in the broadcast call use the same cipher key and the same ciphering algorithm such that mobiles performing handover are able to understand the downlink in the new cell.

The complexity arises in the cipher key management. Obviously the key could be changed by recalling the SIM cards and re-programming them. However, this is too laborious for many users and so over-the-air updating can be performed using the SIM toolkit. To achieve this a point-to-point call is established with each mobile separately and authentication and point-to-point ciphering performed. The new group key, or keys, can then be sent.

The broadcast call originator will be using his own dedicated (point-to-point) ciphering key at the time that the call is established. He retains this key on his dedicated channel throughout the duration of the call.

Paging and notification whilst in a dedicated call

There is a requirement that mobiles are made aware of new group or broadcast calls even when they are within an ongoing point-to-point, group or broadcast call. This could be important if, for example, a broadcast call was used to inform train drivers about an emergency - even those in a point-to-point call should ideally have their call interrupted to receive this message.

This functionality is clearly very useful but is not achieved without some disadvantages. The detailed architectural process whereby this is achieved is described in more detail in Section 5.2. In summary, in order to make mobiles aware of new calls whilst in a broadcast call, information about the new calls must be sent on the control channels relating to the call (the SACCH and FACCH). Most information will be sent on the SACCH, avoiding any disruption to the speech. Some more critical information (indicated by its priority) might be sent on the FACCH, causing a small disruption to speech. Furthermore, the capacity of the SACCH is limited, and with a high number of paging and notification messages the SACCH might become overloaded. In this case, the delay before a mobile becomes aware of another call will rise. A further disadvantage is the increase in complexity in the BSS.

In the future it is hoped that mobiles with dual-receivers will emerge. In this case, the mobiles will be able to monitor the paging and notification channels in addition to listening to the call. The capacity constraints on the associated control channels will then be lifted. In the meantime, operators are free to use the priority of the call to determine how often notifications or paging for that call should be sent, and other means of enhancing the performance through better scheduling could be implemented - considerable freedom has been provided within the specifications in this area.

The use of channels

It is possible to optimise the use of radio resources in a broadcast call in order to reduce the requirement on network resources or to reduce the cost of the call. The network can decide not to allocate (downlink) broadcast channels in some of the cells in the group call area. In the case that resources are not allocated, notification messages in the cell would inform mobiles of the existence of the call but would not provide information giving the location of the channel. Mobiles must respond to notification messages (in the same manner that response to paging messages is made).

In those cells where a response to the notification messages was received, a broadcast channel is allocated on one of the standard traffic channels. Mobiles performing cell re-selection during a broadcast call to a cell where the broadcast channel was not allocated would respond to the notification messages in the new cell relating to the broadcast call they are listening to and a channel would be allocated. This feature also has the additional advantage, as a further option, of allowing authentication of each of the listeners (and indeed may be used explicitly for that purpose).

Another feature is the possibility to periodically check whether, in cells which have a shared channel allocated, there are any mobiles present. This is achieved by sending a message on the downlink in the individual cell asking for any mobiles in that cell to make a random access attempt on the uplink. If random access attempts are received then the shared channel is left in place. Otherwise, the shared channel can be withdrawn and the notification messages changed such that they no longer include the channel description.

The disadvantage of this feature is that it will significantly extend set-up times in some cells and may result in longer handover breaks when moving into cells where the channels have not been allocated. Operators will need to determine their priorities when using this option.

Acknowledgements

Many users require that critical broadcast messages be acknowledged. There is a wide variety of bearers within the GSM specification and users can chose any of these to carry the acknowledgement. Typically, it is thought that the SMS service would provide an appropriate mechanism for sending acknowledgements. The mechanism to use for acknowledgement is not being specified within ASCI. The need for an acknowledgement could be noticed by the mobile in a number of ways, for example based upon the MLPP priority information.

Battery saving measures

In order to reduce battery requirements for mobiles in idle mode listening for group and broadcast calls an optional feature is a form of discontinuous reception (DRX). In each paging subgroup a 2-bit number is transmitted at the end of each message. When a new message is broadcast on the notification channel this number is incremented. When a mobile monitors its paging subchannel, if this number is different from the last time it read the notification channel, the mobile should re-read the notification channel.

Microcells and macrocells

Because a group or broadcast call is required to cover a specific area, smaller cells cannot be used to increase system capacity[1]. Indeed, the coverage method requiring the least resources would be to use a single cell covering the entire group call area. For this reason, some network

[1] Smaller cells provide increased capacity for point-to-point calls because for a given amount of radio spectrum, the number of calls per cell is fixed, but if the cells are made smaller then more cells cover the same area and hence the overall capacity within that area is increased. In the case of broadcast or group calls, if more cells are placed within an area, the call needs to be carried on more channels and hence the additional channels gained from having smaller cells is exactly offset by the need to use more channels to broadcast the call over the same area.

architectures may decide to support point-to-point calls on the smaller cells and group and broadcast calls on the oversailing (umbrella) macrocells. The effect of this is that mobiles in the smaller cell should be notified of calls existing on the larger cell, that is the notification messages should be able to send the mobile to a different cell. This functionality was not added into ASCI as it was felt to complicate the standard but may be provided in later releases, particularly as the GPRS standard provides solutions for similar problems.

3. THE VOICE GROUP CALL SERVICE

3.1 Overview of the service

The voice group call service (VGCS) provides a facility whereby either a mobile user or a fixed user can establish a group call channel to a number of listeners each of whom may then decide to transmit speech to all the other listeners. Basically, the group call is the same as the broadcast call except after the originator has finished speaking other mobiles are able to speak. As with the broadcast service, the listeners are:

- all the mobiles who are members of the group, who are within a pre-defined coverage area and who are able to accept the call;
- all the pre-defined fixed users.

The originator will talk for a time and then become a listener. The originator is not aware of who is listening to the call but can request acknowledgements to be sent after the call is completed. Once there is no talker, any listener can request the uplink and if this request is granted he can speak until they decide to return to being a listener. Either the originator or an authorised listener can terminate the call, or alternatively, the network can terminate the call after a pre-defined period of silence.

3.2 Particular features

Group calls share all the attributes and options of broadcast calls as described above. In addition, they have the extra features described in this section.

Termination of a group call

Termination of a broadcast call is relatively simple, it is either directly terminated by the originator, or if contact is lost with the originator. The issue is not so simple with group calls. Only the originator, a group member with a subscription which allows termination (typically a despatcher) or the network are allowed to terminate the call. If the originator loses contact with the network whilst he is a listener this may not be known to the network so additional means must be possible to terminate the call. A timer is used which terminates the call after a defined period of no voice activity.

Echo cancellation and dual-downlink usage

In the case where the talker in a group call uses the uplink of the common channel, their voice will be re-broadcast on the downlink so it can be received by the other users. To avoid annoying echo effects, the downlink in the mobile is muted when the uplink is in use. However, it is required that certain important group members are able to over-ride the speaker in order to relay important instructions. When such a user interrupts the speaker in a group call the speaker's mobile receives a signal to lift the downlink muting in order for the talker to be aware that they are being interrupted. The speaker will then hear a combination of the other users and an echo of his own voice, which may well be unintelligible. This should be sufficient to persuade the talker to stop so that he can listen to the other user.

Uplink arbitration

Arbitration is provided such that only a single user is allocated the uplink at any given time. This arbitration is normally performed by the MSC which is responsible for the call. However, in the case where very rapid uplink arbitration is required, initial arbitration can be performed at BSC level. If the BSC controlling the cell in which the uplink was requested believes that the uplink was free it allocates it to the user and allows him to start talking. Simultaneously it sends an uplink request to the MSC. If this is accepted no further action is required. If the MSC has, in the meantime, allocated the uplink to another user then it rejects the request from the BSC. The BSC must then remove the uplink entitlement from the mobile.

The loss of the talker

Difficulties can occur if the uplink connection for the talker using the common uplink of a group call is suddenly lost, probably due to propagation problems. Timers are started in the network and the mobile and if contact has not been established after a certain period of time (which can be an operator-defined parameter) the uplink is declared free. If the mobile returns to the call after this timer has expired it must consider itself to be a listening mobile.

The use of shared and dedicated channels

As with broadcast calls, the default position is to allocate a shared channel in each cell where the group call is to take place and to leave the originator on his dedicated channel until he moves to group receive mode. However, substantial flexibility is allowed. Essentially, the network can decide, at any time, to move participants who have dedicated channels to the shared channels, to move the mobile using the shared uplink to a dedicated channel, and to put mobiles responding to notification messages which were provided without a channel description[2] on either a dedicated or a shared channel.

This flexibility may be useful, for example, if it is known that there is only one mobile in a cell (e.g. only one mobile responded to notification messages). Then the mobile could be allocated a dedicated channel which would allow the user to perform other functions on this channel (such as receive SMS messages). Furthermore, the network can use downlink power control on the channel, which it is not able to do on a shared channel, reducing the interference experienced in neighbouring cells.

2 Mobiles do not, of course, respond to a notification message when a full channel description was provided in the notification message, they simply move to the indicated channel and start listening.

4. THE ENHANCED MULTI-LEVEL PRIORITY AND PRE-EMPTION SCHEME

4.1 Overview of the service

The enhanced multi-level priority and pre-emption (EMLPP) scheme provides a means whereby each call can have a priority associated with it. Based on this priority, the network can make decisions to pre-empt ongoing calls if there are insufficient resources or to interrupt a call to a mobile if a new call for that mobile is of sufficiently high priority. The mobile can also make decisions to automatically move to a new incoming call when in an existing call if the priority of the new call is sufficiently high. This mobile functionality is of use when the mobile is in a group or broadcast call such that the network does not have a dedicated connection with the mobile but the mobile is not in idle mode.

Priorities within GSM are aligned with those within ISDN. The ISDN service is known as multi-level priority and pre-emption (MLPP) and the GSM service has been termed enhanced MLPP (EMLPP). Within GSM, seven priority levels are provided whereas MLPP, as provided in ISDN networks, only allows five priority levels. The MLPP priorities are aligned with the lowest five EMLPP priorities. If a GSM call passes through an ISDN network, the highest three priorities are mapped onto the highest MLPP priority.

A priority is specified at the start of a call, either by the originator or a default is supplied by the network. At each point in the network where congestion could occur (e.g. Um interface, A-interface, E-interface) the entity responsible for allocating the resource determines whether the resource is available. If it is not, this entity considers the priority of the calls using the resources and determines, based upon a lookup table, which, if any, of the calls should be pre-empted or queued and performs the appropriate activity. At all points in the network the only priority to apply is that specified by the originator. The priority subscription of the called party is not taken into account at any point.

In the case of group and broadcast calls, only a default priority is applied. This priority is set within the group call register for each group and each service area. This is sensible as, for example, when making a call to an emergency group, only an emergency priority should apply.

4.2 Particular features

Default priority

The default priority is held in the network along with other subscription information. It is changed by the service provider. It is likely that the level of priority will be related to the subscription charge. In the case that the mobile does not specify a priority when establishing a call, the default priority in their subscription is applied. The level of priority selected is then signalled back to the mobile so that it is able to make pre-emption decisions should there be other incoming calls.

5. DETAILED ARCHITECTURAL DECISIONS

In this section, some of the detailed decisions made during the standardisation of the ASCI features are described. This provides an enhanced understanding of the structure and methodology of the ASCI features.

5.1 The channel type

One of the first choices was to decide what form of channel would be used for group and broadcast calls. The options and advantages and disadvantages of each are discussed below:

Channel per mobile. In this approach a separate channel is provided for each mobile in a group or broadcast call. This would have provided essentially the same service as the current conference call but perhaps with more automated call establishment procedures. Its advantages were ease of implementation but it had the major disadvantage of inefficient use of radio resources and was rejected for this reason.

Dedicated channel per cell. In this realisation, a single channel is permanently assigned in each cell for group or broadcast calls. When entering the cell, the mobile monitors the channel for activity and if it finds any it listens to the group call. The advantages are simplicity through dedicating resources and the avoidance of the need to notify the mobile of the group call. However, there are many disadvantages including inefficient use of resources, difficulties if there is more than one group call per cell and difficulties for the mobile to know whether the call relates to a group he is a member of. The solution was rejected for these reasons.

Single channel per cell. In this approach, which was the one adopted, a single channel is provided for each on-going group call in each cell where coverage is required for the duration of that call. If the talker is in that cell, he uses the uplink of this channel to relay his speech.

Dedicated channel per cell plus channel per talker (the "one and a half channel" solution). In this solution a group channel is assigned in each cell for each on-going group call for the downlink transmission and a separate channel is assigned to the current talker. This is permanently assigned to the originator in the case of a broadcast call, and assigned on demand to talkers in the case of a group call. This solution offered the advantages of simpler implementation as the talker was treated as a standard point-to-point call and the group channel was a simple downlink. The disadvantages were slightly greater resource requirements and the need to dynamically link channels into the group call bridge as a new channel was assigned to a talker. Initially the one and a half channel solution was rejected. However, when considering call initiation it became apparent that a group call initiator would signal on a dedicated channel and then need to move to a group call channel. Because all the resources had already been allocated and because of the difficulties in moving the originator during the early stages of the call, it was decided that the originator should keep his dedicated channel until the network was happy to move him to the common channel. In this solution the originator of a broadcast call will permanently keep a dedicated channel. The originator of a group call may keep a dedicated channel until he stops talking or he may be required by the network to transfer at an earlier point.

Subsequently, it was decided that the network should have the flexibility to allocate dedicated channels to some users either in addition to, or as an alternative to, the group channel. For example, if it is known that there is only one user in a particular cell then the use of a dedicated channel in this cell for the group call offers a number of advantages.

The strategy of providing a group channel in all the cells within the required coverage area makes inefficient use of resources in the situation where there are only a few users in the coverage area as there may be cells with no users in. A group call channel will be established in these cells even though it is never listened to. For group calls where a rapid call establishment is not required, mobiles could be made to respond to notification messages. Only in those cells where a response was received would a common channel be allocated. Mobiles performing cell re-selection during a group call would signal when they moved into a cell not currently supporting the group call and the network could make a decision as to whether a group call channel should be allocated in this cell.

5.2 Paging, notification and late entry

The requirements for group and broadcast calls stated that mobiles should be made aware of new group or broadcast calls even when they were within ongoing point to point, group or broadcast calls. Furthermore, when moving into an area where there were ongoing group or broadcast calls they should be made aware of these, even when in an on-going point to point, group or broadcast call. These requirements were onerous and caused much discussion as to how they could be met.

Initially it was intended that mobiles would read the paging channels when in a group or broadcast call (and ideally when in a point-to-point call as well). However, this is only possible if the mobile has a dual-receiver with one receiver dedicated to the call and the other to monitoring the control channels. Even with two receivers there would be situations where not all messages could be received, but further work would probably have discovered solutions to this. Unfortunately, the dual-mobile solution could not be recommended as mobile manufacturers were not prepared to make such mobiles. In the long term, dual-receiver mobiles will be available as services such as high speed circuit switched data may require them. At this point the ASCI service can be simplified. Until such a time, alternative solutions were devised which would provide acceptable functionality.

Since a mobile in a call can only monitor one channel it was apparent that all signalling would need to be relayed down this channel. The most suitable place for the signalling was the SACCH. There then followed much discussion as to whether the SACCH offered sufficient capacity for paging and notification and the inefficiencies caused by such repetition of signalling. Both these concerns were valid, but no alternatives presented themselves and so the decision was taken to send paging and notification on the SACCH or the FACCH if the latter seemed more appropriate. Possibilities have been provided for a network operator to tailor the notification process to his own particular traffic profile and needs (e.g. in relation to EMLPP).

All the possible cases are shown in the table below, with the solutions for each case elaborated where necessary.

	Current mobile status		
New call	*Idle*	*point-to-point*	*group/ broadcast*
point-to-point	Paging	Call waiting	SACCH/ FACCH (3)
group/ broadcast	Notification	SACCH (1)	SACCH/ FACCH (4)
late entrant	Notification	FACCH (2)	FACCH (5)

1) Notification messages for group and broadcast calls within the cell are relayed on the SACCH into point-to-point calls. Optionally, if the mobile has indicated by its classmark that it is not able to support group and broadcast calls this notification will not be performed.

2) When entering a new cell where there are on-going group and broadcast calls, details of all on-going group and broadcast calls can be provided on the FACCH. This quickly informs mobiles of the calls without significant interruption.

3) Typically, the paging load will be too great to relay in its entirety on the SACCH or FACCH without unduly affecting call quality and so instead the number of any paging subchannels which have new paging messages are sent on the SACCH. Mobiles in these paging subgroups can then listen to the paging channels with an increased probability that there

will be a call relating to them. When they listen to the paging subchannel there will be a short break in the received speech. Generally, these breaks will be sufficiently short and infrequent to be hardly noticed by the user.

4) All notification for new and on-going group and broadcast calls must be repeated into all on-going group and broadcast calls. As an enhancement, a number can be broadcast on the SACCH of the group or broadcast call. If there is a new notification message, this number is changed. The mobiles recognise that there has been a new message added to the notification channel and listen to the notification channel to read the new message. This reduces the information sent on the SACCH but results in an occasional short break as the mobile reads the notification message.

5) As (2).

5.3 SCCP connections

An area which has caused much difficulty and discussion during the standardisation has been that of terrestrial circuits allocated. In principle, only one circuit is required between each MSC in the group call and between each MSC and each BSC. However, such an approach raised a number of difficulties and caused an increase in the standardisation work required. The main difficulty was associated with the transcoder and rate adaptation unit (TRAU). Since most implementations have the TRAU at the MSC site and since the BTSs are not synchronised, using a single circuit would cause severe difficulties. As a result it was decided that a circuit per cell would be allocated between the MSC and the BSC.

The picture was not so clear between MSCs. Here there is no major technical obstacle preventing the use of a single circuit. In addition, MSCs are rarely co-located and the amount of traffic flowing on the E-interface has revenue implications for the operators. For that reason, it was decided to only use a single channel per group or broadcast call on the E-interface. The implication of this is that a secondary conference bridge is required in the other MSCs. This bridge distributes the signal from the originating MSC to all its BSCs required and takes the uplink signal from a BSC if the uplink has been allocated in that BSC and returns this to the originating MSC. As such, the functionality of this secondary bridge is limited and does not add significant functionality to the system.

5.4 Division of functionality

The anchor MSC

It was originally envisaged that the main, or anchor MSC for a group or broadcast call would be the MSC containing the cell from within which the group or broadcast call was originated. This offers a straightforward implementation and seemed sensible. However, a somewhat esoteric problem was then encountered. There is a requirement that no two identical group or broadcast calls (i.e. for the same area and same group identity) be ongoing at the same time. Otherwise, it could not be guaranteed which mobiles were in which calls and mobiles could become very confused by notification messages. In order to prevent this, the GCR contains a flag for each group and service area (the combination of these two is termed a group reference) indicating whether a call is ongoing for that call reference. The anchor MSC consults this GCR at the point of call origination to determine whether a call is ongoing and if so rejects the call origination attempt. So far, so good.

Now it may happen that a group call area covers more than one MSC and that the identical group call is required (i.e. the same coverage area results) whether the call is established from either MSC. The MSCs may also be linked to different GCRs. Now checking for an ongoing call becomes more difficult. This record could be stored in both GCRs, but if two simultaneous call requests are received in different MSCs there is no deterministic way of arbitrating between them. The only solution is to nominate an anchor MSC for each possible group and broadcast call. In this case, the anchor MSC may not be the one from within which the call was originated which is a divergence from normal GSM practice. This is not a major problem, simply requiring a message from the originating MSC to the anchor MSC shortly followed by a message back to the originating MSC which now becomes a relay MSC.

Organisation of notification channels

It was decided to create a new channel called the Notification Channel NCH. In order not to disrupt the paging functionality in present networks the NCH is situated inside the Access Grant Channel (AGCH), thus making the AGCH a mandatory feature if VBS/VGCS features are to be supported. The number of blocks to be used for notification and their exact position can be found on the BCCH.

In order to reduce battery requirements for mobiles listening for group calls an optional feature is a form of DRX. In each paging subgroup is a 2-bit number. When a new message is broadcast on the notification channel this number is incremented. When a mobile monitors its paging subchannel, if this number is different from the number the last time it read the notification channel the mobile should re-read the notification channel.

5.5 The GCR

A new entity has been proposed to hold the information relating to the group and broadcast calls. This has been termed the group call register (GCR). This contains for each possible group:

- details of the anchor (see below), and if the anchor the following information:
- the area over which the call is to be made;
- a list of despatchers to be included in the call;
- the default priority of the call;
- the time-out period for the call in the case of no speaker activity;
- a flag to indicate whether the call is ongoing.

The GCR is connected to the MSC. Few restrictions have been placed on GCR deployment. It is not foreseen that there would be more than one per MSC, but there may be a single GCR for more than one MSC. It was decided not to standardise the interface between the GCR and the MSC. This was simply because nobody offered to perform the necessary work and this interface is a possibility for future standardisation.

ACKNOWLEDGEMENTS

Standardisation of features within ETSI is a process into which many people contribute. Over 40 individuals took part in the ASCI standardisation at one time or another. Those who took on particularly significant parts of the ASCI specification work include Ansgar Bergmann, Roland Bodin, Michel Mouly, Dirk Munning and Michael Roberts whilst Peter Van de Arend provided excellent secretarial support. Within the UIC, a group of people led by Michael Watkins and including Les Giles and Jos Nooigen worked hard on the specifications whilst Richard Shenton from Smith System Engineering did much to help the smooth progress of the project. Additional thanks are due to Michael Roberts for his careful correction of the manuscript.

REFERENCES

[Webb] W T Webb and R D Shenton, "Railway communications", IEE Electronics and Communications Journal, Vol.6, No.4, Aug 1994, pp195-202.

Chapter 3

General Packet Radio Service
Standards Overview

JARI HÄMÄLÄINEN
Nokia Mobile Phones

Key words: GPRS, Packet Radio, GSM Data, Wireless Data.

1. INTRODUCTION

General Packet Radio Service (GPRS) belongs to GSM phase 2+. GPRS is a data service that provides a packet radio access for GSM mobile stations (MS) and a packet switched routing functionality in GSM infrastructure. Packet switched technology is introduced to optimise bursty data transfer and occasional transmission of large volumes of data. The idea of GPRS was first time discussed during 1992 and the first release of the standards is targeted to the end of 1997. That release covers all the major functionalities of GPRS, including point-to-point transfer of user data, Internet and X.25 interworking, quick SMS transfer using GPRS protocols, filtering functionality for security reasons, volume based charging tools, anonymous access that can be used e.g., for prepaid cards in the road traffic information systems, and roaming between public land mobile networks (PLMN). The second release will be published one year later and it will include point-to-multipoint (PTM) transfer (PTM-Group call and PTM-Multicast), supplementary services, and additional interworking functionality (e.g., ISDN and modem interworking).

For users, the most important advantage of GPRS is the possibility for charging that is based on traffic volume. There is no need to pay for unused transmission capacity. During idle periods, the spectrum is given effectively to other users. Another main issue is that GPRS is a proper carrier for most of the existing, as well as for new communications applications, because the system provides a variable transmission capacity with maximum data rate up to 171.2 kbps.

From the network operators point of view, the scarce system resources must be used efficiently. Especially in data services, the burstiness of transmission enables sharing of the radio interface and network resources by several users, without decreasing the efficiency of each individual transmission too much. That enables increased income per unit of available spectrum, even if each user experiences decreased operational expenses.

The service is targeted mainly at applications with traffic characteristics of frequent transmission of small volumes and infrequent transmission of small or medium volumes. This enables the system to attract new services and applications. The transmission of large volumes should still remain via circuit switched channels, to prevent blocking the packet radio spectrum. The possible applications for GPRS range from communication tools in a laptop PC (electronic mail, file transfer, and WWW browsing) to special applications with relatively low transmission needs (telemetry, road and railway traffic control, taxi and vehicle dispatch, dynamic road guidance, and monetary transactions).

GPRS service can be used with standard software protocol packages. The interface between GPRS protocol stack and the application protocols is based on Point-to-Point Protocol (PPP) or some commonly used driver, e.g., Network Driver Interface Specification (NDIS).

2. GPRS NETWORK ARCHITECTURE AND PROTOCOLS

GPRS network architecture is built on top of the existing GSM network infrastructure. However, several new network elements are needed for the packet switched functionality (see Figure 1). The main routing functionality is carried out by support nodes. There exist a Gateway GPRS Support Node (GGSN) and a Serving GPRS Support Node (SGSN). Additionally, there is a backbone network that connects the SGSN and GGSN nodes together, and a border gateway that handles the packet transfer between GPRS PLMNs. A domain name server can be used for address translation purposes.

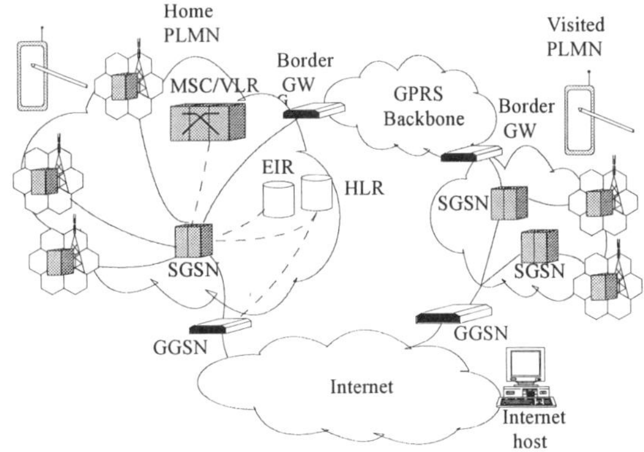

Figure 1. Logical Architecture for GPRS PLMNs

GGSN maintains the location information of the mobile stations that are using the data protocols provided by that GGSN. It is also an interworking node between GPRS PLMN and the external data network (i.e., Internet or X.25 network). Based on the address of a packet received from the external data network, the GGSN is capable of tunneling the packet to an appropriate SGSN. Also, the packets transmitted by a mobile station via SGSN to GGSN are routed to the external data network. The routing of data packets is possible since the GGSN is participating into the mobility management procedures of GPRS. An important feature for the network operator is that GGSN is capable of collecting charging information for billing purposes.

SGSN participates into routing, as well as mobility management functions. It detects and registers new GPRS mobile stations located in its service area and transfers data packets between mobile stations and GGSNs. SGSN operates the high level radio interface protocols, as well as numerous GPRS network protocols.

Tunneling of data and signalling messages between GPRS support nodes is carried over a GPRS backbone network. The protocol architecture of the Backbone network is based on the Internet Protocol (IP). For network protocols requiring reliable transfer over the GPRS backbone (e.g., X.25), Transmission Control Protocol (TCP) is used with IP. Otherwise, the User Datagram Protocol (UDP) is used with IP (e.g., for Internet communication). On top of the previously mentioned protocols, there is a GPRS Tunneling Protocol (GTP). When data transfer between two GPRS PLMNs is needed, a border gateway is used to provide appropriate security over the backbone network. The type of the

Backbone network, that is selected by a roaming agreement, can be a public Internet or a leased line.

The operation of low level radio interface protocols is taken care by the Base Station Subsystem (BSS). The medium access and automatic re-transmission protocols are the main functions of GPRS base station subsystem. Since there exists large number of base station subsystems in operation, the GPRS protocols are designed so that the existing equipment can be upgraded for GPRS usage.

In case the control of the radio interface is not taken care by the base transceiver station, a remote Packet Control Unit (PCU) may be implemented. In that case the radio control functions locate remotely, in Base Station Controller (BSC) or SGSN site. The transfer of data and signalling messages between base transceiver station and PCU is carried out using PCU frames that are extended Transcoder Unit (TRAU) frames.

The protocol architecture between base station subsystem and SGSN is based on the Frame Relay, utilising the virtual circuits to multiplex data from several mobile stations. The link may be point-to-point or multi-hop. A GPRS specific BSSGP protocol (base station subsystem GPRS protocol) is used on top of the Frame Relay. BSSGP provides message formats and procedures for transfer of data and paging messages, and mechanisms for management of the link. The protocol architecture of GPRS transmission plane is described in Figure 2.

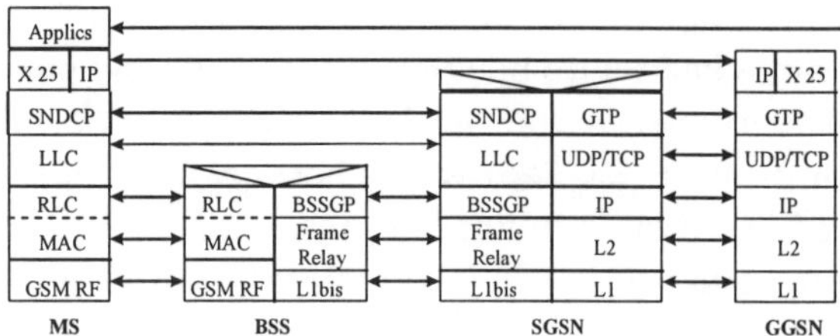

Figure 2. Protocol architecture in GPRS transmission plane

Mobile Switching Center and Visitor Location Register (MSC/VLR) are not needed for routing of GPRS data. However, the MSC/VLR is needed for the co-operation between GPRS and the other GSM services.

Home Location Register (HLR) contains the GPRS subscription and routing information. Authentication Center (AuC) takes care of the generation of authentication, as well as ciphering parameters. Equipment Identity Register (EIR) is used for authentication of the mobile equipment, e.g., to enable removal of the stolen mobile equipment from the network. SGSN interacts with MSC/VLR, HLR, AuC, and EIR, using the Signalling System #7 protocols. The same applies to interaction with SMS-gateway MSC, or SMS-interworking MSC.

3. GPRS MOBILITY MANAGEMENT

Layer three Mobility management (L3MM) protocols are used to support continuous service, independently of the mobility of the subscriber. In GPRS, the mobility management functions cover the initialisation of the service and packet data protocol (PDP) context, as well as the tracking of the location of the subscriber.

3.1 Mobility Management States

There exists three mobility management (MM) states in GPRS, i.e., Idle, Standby, and Ready. In the Idle state, the mobile station may perform PLMN selection, GPRS cell selection, and re-selection. However, the mobility management and routing contexts are not active in the mobile station and the SGSN. The mobile station may receive only point to multipoint - multicast (PTM-M) data.

Standby is a state where mobile stations are normally prepared for data transfer, but are not actively transferring at the very moment. In Standby state, the mobility management context between mobile station and the SGSN is active. The mobile station informs SGSN every time it changes from one routing area to another. The routing area is a set of cells (from one cell up to the size of a location area) defined by the operator. The mobile station may receive pagings for circuit switched services, as well as pagings for GPRS point-to-point (PTP) and point-to-multipoint group call (PTM-G) data. Additionally, reception of PTM-M data is possible.

When the mobile station is willing to send or receive data (except PTM-M), it must enter into the Ready state. In Ready state, reception of data is possible without paging procedure, because the network knows the location of mobile station in the accuracy of one cell. Mobile station informs SGSN every time it switches between cells. Ready state is guarded by a timer. The timer is reset each time after reception or transmission of a packet. When the timer elapses, the mobile station enters into the Standby state. Change from Standby to Ready state may be initiated by the network, using a paging procedure. That is used when there is data to be send to the mobile station. When the mobile station has data to send, it may initiate the transfer immediately and the state will be changed automatically from Standby to Ready.

3.2 Attach Procedure

When GPRS subscriber wants to send and receive data, an Attach procedure is carried out. During the Attach procedure (GPRS Attach, IMSI Attach or combined GPRS/IMSI Attach), the mobile station enters into the Ready state. The Attach procedure is initiated by the mobile station. In the Attach procedure, the identity of the mobile station, ciphering key sequence number, and classmark of the mobile station are delivered to the network. The identity of the mobile station is often a temporary logical link identity (TLLI) combined with the routing area identity identifying the routing area where the TLLI has been given. In case the mobile station has never had TLLI before, IMSI may be used. If the identity is known only by another SGSN, an identification from the other SGSN is queried. The SGSN may requests authentication of the mobile station, the ciphering may be initialised, and the IMEI of the mobile station may be checked. If necessary, the location information in HLR is updated, the context in the old SGSN is deleted, and the subscription information in the SGSN is updated. Also, the location information is delivered to the new MSC/VLR and the information in the old MSC/VLR is deleted. It is possible to change the TLLI during the Attach procedure. Once the Attach procedure is completed, the mobile station may transfer short messages (SMS), receive PTM-M messages, or activate a PDP context for some packet data protocol.

3.3 Activation of PDP Context

A natural continuation for the Attach procedure is activation of PDP context. Normally, the mobile station requests the network to activate the desired PDP context with requested quality of service. However, the PDP context activation may be requested also by the network. In PDP context activation, the routing context in GGSN is activated. Routing between SGSN and GGSN is enabled by the activation of tunneling identifier in the SGSN and GGSN. PDP context may be activated for fixed, as well as dynamic address. After GPRS Attach and activation of one or more PDP contexts, transfer of point-to-to-point and point-to-multipoint data is possible.

3.4 Informing the Location of Mobile Station

When the GPRS mobile station is moving, the location must be known by the network. In Ready state, the mobile station informs its location to SGSN every time it enters into a new cell. All the uplink data or signalling messages are valid cell updates. However, when the mobile station enters into a new routing area, a routing area update procedure must be initiated. In that procedure, the routing area information is updated in the SGSN. The procedure is mandatory for the mobile stations in Ready, as well as in Standby states. If the serving SGSN is not the same as the old SGSN, the mobility management and PDP contexts are fetched from the old SGSN. The GGSN is informed about the new SGSN. Additionally, the location information in HLR is updated. HLR requests the old SGSN to cancel the location information for that mobile station. The location information in the MSC/VLR must be updated, if the mobile station is also IMSI attached. The routing area and the location update procedures may be combined, if the mobile station is GPRS and IMSI attached.

4. GPRS RADIO INTERFACE

GPRS defines a set of new logical radio channels. Packet Common Control Channel (PCCCH) is a set of logical channels used for common control signalling. PCCCH consists of:

- Packet Random Access Channel (PRACH) used by mobile station to request access for uplink transfer
- Packet Paging Channel (PPCH) used by the network to inform mobile station about downlink transfer
- Packet Access Grant Channel (PAGCH) used by the network to reserve a packet traffic channel for mobile station
- Packet Notification Channel (PNCH) user by the network to notify a group of mobile stations prior to PTM-M packet transfer.

Packet Broadcast Control Channel (PBCCH) is used by the network to send system information messages for GPRS mobile stations.

Packet traffic channels are dedicated temporarily for mobile stations. Packet Data Traffic Channel (PDTCH) is used for data transfer. Data may be point-to-point or point-to-multipoint messages or GPRS mobility management messages. Low level signalling related to a given mobile station is transmitted on a Packet Associated Control Channel (PACCH). Such messages include acknowledgements of data, resource allocations, and exchange of power control information.

GPRS follows a 52-multiframe structure. In 52-multiframe, every 13th TDMA frame is idle. The idle frames are used by the mobile station for the identification of the base station identity codes, continuous timing advance update procedure, and interference measurements for power control purposes. Rest of the frames are used for the GPRS logical channels. Re-use of the 51-multiframe structure for PCCCH is also allowed by the standard.

The protocol hierarchy on the radio interface is drawn in Figure 3. The description of each protocol is given in the following text.

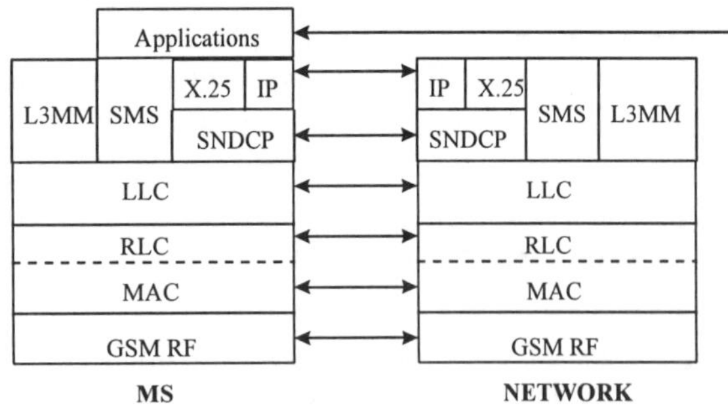

Figure 3. GPRS protocol architecture between mobile station and SGSN

4.1 Subnetwork Dependent Convergence Protocol

Between mobile station and SGSN, a network protocol data unit is segmented by Subnetwork Dependent Convergence Protocol (SNDCP) into one or several Logical Link Control (LLC) frames. In addition to the segmentation function, the SNDC protocol includes multiplexing of user data, user data compression, TCP/IP header compression, as well as transmission based on the requested quality of service (QoS).

A variety of network layer protocols may be carried by the SNDCP. Such a set includes IP, X.25, PTM-M, and PTM-G. The network layer protocols are separated by a network service access point.

The V.42bis algorithm is used for compression of user data. Additionally, compression of TCP/IP headers may be applied. An independent compression procedure is performed for each of the QoS classes.

4.2 Logical Link Control Protocol

LLC is a protocol that provides a reliable logical link between mobile station and SGSN. The SNDCP, SMS, and GPRS mobility management messages are transmitted in the LLC frames. The LLC frame includes a frame header with numbering and temporary addressing fields, variable length information field, and a frame check sequence. The maximum length of the LLC frame is almost 1600 octets. However, the mobile station and the network may negotiate a shorter maximum frame length for the logical link. The functionality of LLC includes maintenance of the communication context between the mobile station and the SGSN, transmission of acknowledged and unacknowledged frames, detection and retransmission of corrupted frames. LLC frames are transmitted in one or several radio blocks. The logical link is maintained when the mobile station moves between cells within the service area of one SGSN. However, when the mobile station move into a cell that is served by another SGSN, a new logical link must be established.

GPRS ciphering is operated at the LLC layer. The ciphering key is achieved during the Attach procedure. A new ciphering algorithm that is appropriate for variable length frames is being developed for GPRS.

4.3 Medium Access Control and Radio Link Control

The Medium Access Control (MAC) is used to share the radio channels between mobile stations, and to allocate the physical radio channel for a mobile station when needed for transmission or reception. The MAC functionality uses the GPRS logical channels, or alternatively, the existing GSM logical channels.

When the mobile station has a packet to transmit, a packet channel request is transmitted. In case of congestion, or loss of the packet channel request, a quality of service specific back-off algorithm is applied to the scheduling of the retransmission of the packet channel request. The access may proceed in one or two phases. In the one phase approach, the packet channel request message contains all the information needed for establishment of the channel, e.g., multislot related information and quality of the requested service. The base station subsystem acknowledges the request by sending a packet immediate assignment message containing the information about the physical channels and temporary flow identity reserved for the mobile station. From now on, the mobile station may start data transmission on the allocated packet data channels.

In two phase access, the packet channel request leads to reservation of a packet associated control channel. The base station subsystem allocates the control channel where the mobile station may transmit a packet resource request containing all the details of the requested service. A packet resource assignment message containing the information about the physical channels reserved for the mobile station is used as an acknowledgement for the packet resource request.

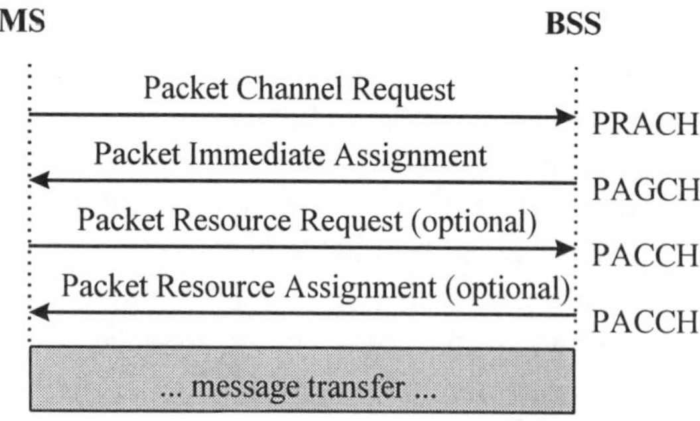

Figure 4. One and two phase medium access for uplink message transfer

Packet transfer from the network to mobile station is initiated by the packet paging request message on a packet paging channel (PPCH). Mobile specific discontinuous reception periodicity may be applied for the paging, to enable optimisations in power consumption. The mobile station indicates its transition to Ready state by proceeding the uplink access for sending an LLC frame to the SGSN.

When the mobile station is in Ready state, SGSN forwards the data to the base station subsystem that sends the packet resource assignment to the mobile station. Discontinuous reception may be applied also to the reception of the packet resource assignment. The packet resource assignment includes all the details of the reserved physical channel on the downlink, allowing the mobile station to start receiving data in the frame indicated in the packet resource assignment message.

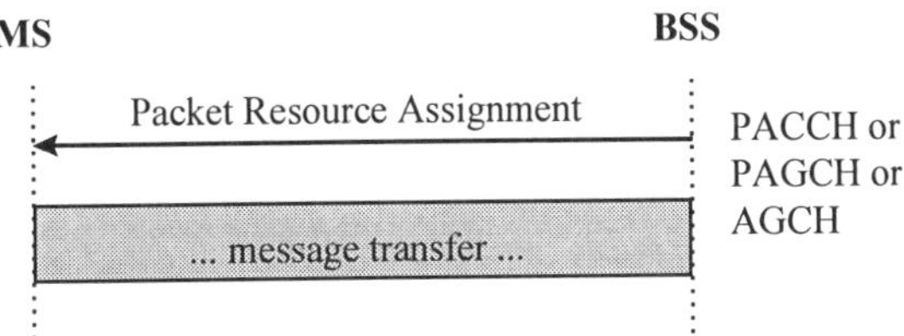

Figure 5. Initiation of downlink data transfer for mobile station in Ready state

The functionality of the Radio Link Control (RLC) between the mobile station and the base station subsystem is to detect the corrupted RLC blocks and request a selective retransmission of the corrupted blocks. The retransmission request is an acknowledgement consisting of a bitmap that indicates each of the received RLC blocks either as corrupted or successfully received. After the acknowledgement, retransmission of the corrupted blocks is done by the transmitter.

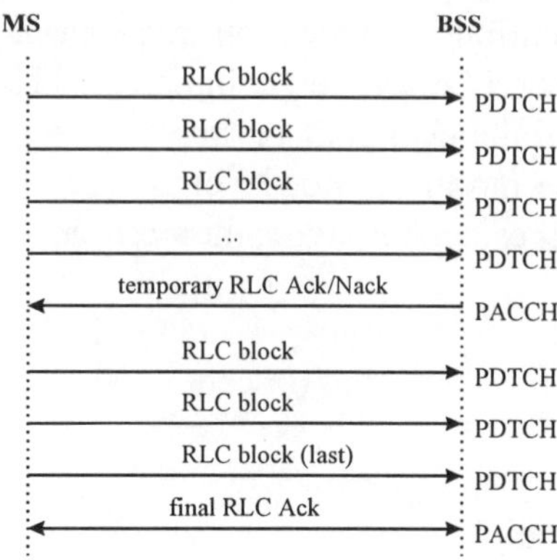

Figure 6. An example RLC data flow for uplink transmission

4.4 Physical Layer

The messages on the radio interface are carried by radio blocks. Each radio block is interleaved over four normal bursts. The radio block consists of a header for media access control, information part used for data or signalling, and a block check sequence. There are four alternative channel coding schemes, i.e., CS-1 to CS-4.

Table 1. Data Rates for GPRS Coding Schemes

Coding Scheme	Data Rate
CS-1	9.05 kbps
CS-2	13.4 kbps
CS-3	15.6 kbps
CS-4	21.4 kbps

CS-1 is the ½ rate convolutional code used for GSM SDCCH channel. In CS-2 and CS-3, there is puncturing applied to CS-1. CS-4 has no forward error correction. High speed packet transmission is enabled by using multiple packet data traffic channels for data transmission. That way the maximum data rate increases up to 171.2 kbps, using all the eight channels of a radio carrier. However, the capabilities of the network or the multislot capability of the mobile station may define a lower maximum data rate. The multislot capability defines also some procedural differences to mobile stations, e.g., full-duplex and half-duplex modes of operation.

Mobile stations shall perform radio signal strength measurements of the signals from the serving and neighbouring cells for cell re-selection purposes. A hierarchical cell structure may be supported by setting a separate priority for each hierarchical level. Based on the measurements, the mobile station shall select highest ranked cell from the highest priority class, provided that minimum signal strength threshold is fulfilled for that priority class. Alternatively, the network may order the mobile stations to suspend the normal cell re-selection procedure, and accept decisions from the network.

Determination of a timing advance value is necessary for the uplink transmission of radio blocks. The network shall estimate an initial timing advance from the packet channel request message. This value shall be used for the uplink transfer until there is need to update it. During the uplink transfer, the mobile station shall periodically transfer access bursts with a predefined contents. The access bursts shall be transmitted during the predefined idle frames in the multiframe structure. The network updates the timing advance value according to the access bursts and delivers the updated value for mobile station.

The spectrum efficiency and power consumption are optimised by power control procedures. The uplink power control algorithm uses a combined open/closed loop power control method. The mobile station listens the power control parameters given by the network in broadcast or channel assignment messages and measures the signal strength from the broadcast channels (when not transferring) or packet data channels (during transfer). Based on these parameters and measurements, the mobile station evaluates whether a specific power level or the maximum allowed power level shall be used. The uplink power control algorithm is given in formula (1).

$$P_{CH} = \min(\Gamma_{CH} - \alpha\, C, P_{Max}) \quad (1)$$

where

Γ_{CH} is a mobile station and channel specific power control parameter. It is sent to the mobile station in any resource assignment message. At any time during a packet transfer, a new Γ_{CH} value can be sent to the mobile station.

$\alpha \in [0,1]$ is a system parameter for which a default value is broadcast on the Packet Broadcast Control Channel (PBCCH). Mobile station or channel specific values may be sent to the mobile station together with Γ_{CH}.

C is the received signal level at the mobile station.
P_{Max} is the maximum allowed output power in the cell.

The downlink transmission power is controlled by the base station subsystem and therefore it is possible to use an operator dependent algorithm. The mobile station shall do interference signal strength measurements that are reported to the base station subsystem. While transferring data, the measurements are done during the idle frames of the multiframe structure. While not transferring data, the measurements are done on the channel defined by the network. A filtered measurement report is transmitted to the network to give the required information for downlink power level.

5. GPRS DATA CONNECTIVITY

Previous sections introduced the basic functionality and operation of the GPRS network, i.e., GPRS bearer. However, for the reader it might be still unclear how to run applications on top of GPRS. From user's and application protocols' point of view, GPRS is one alternative bearer for data transfer.

AT commands defined for GSM are extended with GPRS specific commands. At the beginning of a GPRS session, the transmission mode may be selected with the AT commands. The mode selection includes the protocol to be used and user preferences for quality of service class, packet size, bandwidth, and compression. A GPRS specific script or application is needed to set up the configuration.

A major issue is that there is no need to have any GPRS specific standards for application programming interfaces and protocol driver. The existing interfaces and drivers can be used. When operating Internet Protocol Suite, the PPP can be used to map data communication protocols to the GPRS bearer. The PPP link and network configuration procedures are mapped to the GPRS attach and PDP context activation procedures. Alternatively, the manufacturer of a GPRS mobile station may implement any other driver between the communication protocols and the GPRS protocol stack. One potential interface, because of its wide spread usage, is Network Driver Interface Specification (NDIS).

When operating Packet Switched Public Data Network (PSPDN), the Packet Assambler/Disassembler (PAD) functionality is located in the mobile terminal. The interface between the mobile terminal and the terminal equipment follows the ITU-T X.28/X.29 specifications. The GPRS bearer is used to carry the X.25 packets from the PAD towards the X.25 network, and vice versa.

When the GPRS mobile station is e.g., a communicator, the physical interfaces are manufacturer dependent. However, in case there is a separate GPRS mobile terminal and a terminal equipment, e.g., PC, there is need for standardised physical interfaces. GPRS operates with ITU-T V.24/V.28, PCMCIA, and IrDA interfaces. Any forthcoming interfaces should not be excluded.

6. EFFECTS TO MOBILE STATIONS

The complexity of the hardware for GPRS mobile station is defined by the mobile station multislot class parameter that is used for HSCSD, as well. In GPRS, the number of classes is extended to cover the half-duplex/asymmetric functionality more thoroughly. The standards will allow half-duplex/asymmetric classes that allow 6 or 8 time slots in the downlink while the uplink operates with lower number of time slots. This is possible since the GPRS radio protocols allow flexible scheduling between uplink and downlink transmission.

The processing requirements for the mobile station increase when the number of time slots is increased. Another implication of the multi-slot data transmission is the increase in the power consumption while the mobile station is transferring data. In addition to the increase in the requirements for processing capacity and power consumption, implementation of the new protocols increase the amount of memory needed for program code and buffering. Most of the GPRS radio protocols are new, e.g., L3MM, SNDCP, LLC, RLC, and part of the physical layer protocols. Only some of the physical layer procedures are re-used from the circuit switched GSM services.

7. EXAMPLE OF GPRS DATA TRANSFER

When the mobile station has been successfully attached to the network and the PDP context for IP (in this example) has been established, the mobile station is ready for receiving and transmitting data.

When transmitting data, the terminal equipment passes the user data for GPRS mobile terminal. The IP datagram is compressed and encapsulated into an SNDC protocol unit that is sent, using the LLC, RLC/MAC and physical protocols, over the radio interface to the SGSN that is currently serving the mobile station. When the SGSN has received the packet error-free, it tunnels the packet through the GPRS backbone network to the GGSN that is handling the traffic from the mobile station to Internet. The GGSN removes the tunneling and forwards the IP datagram to Internet that passes it to the recipient.

A host connected to Internet do not need any special skills to communicate with a GPRS mobile station. The host transmits IP datagram through Internet using the IP address for the GPRS mobile station. Internet routing protocols are used to route the packet to the sub-network from which the IP address to the mobile station has been acquired. When the IP datagram arrives at GGSN, the IP address of the receiver is extracted and mapped to the current location of the mobile station. The IP datagram is tunneled through the backbone network to the SGSN that is currently serving the mobile station. The SGSN removes the tunneling and the original IP datagram is encapsulated into the SNDC protocol data unit and sent over the radio interface to the mobile station, using the LLC, RLC/MAC and physical protocols.

ACKNOWLEDGMENTS

The author gratefully acknowledge colleagues from Nokia for a great effort in research and standardisation of GPRS. The GPRS colleagues from ETSI, especially members of GPRS project team (PT80V), deserve thanks for making the standards for a great system real.

REFERENCES

European Telecommunications Standards Institute, GSM 02.60, General Packet Radio Service (GPRS), Service description, Stage 1.

European Telecommunications Standards Institute, GSM 03.60, General Packet Radio Service (GPRS), Service description, Stage 2.

European Telecommunications Standards Institute, GSM 03.64, Overall description of the General Packet Radio Service (GPRS) Radio Interface.

European Telecommunications Standards Institute, GSM 10.60, General Packet Radio Service (GPRS), Project scheduling and open issues.

J. Hämäläinen, Design of GSM High Speed Data Services, Doctor of Technology Thesis, Tampere University of Technology Publications 186, October 1996.

Chapter 4

High Speed Circuit Switched Data

Standards Overview

JARI HÄMÄLÄINEN
Nokia Mobile Phones

Key words: HSCSD, High Speed Data, GSM Data, Wireless Data.

1. INTRODUCTION

Current GSM data services are based on circuit switched technology with a maximum user rate of 9.6 kbps. That rate is applicable for transparent and non-transparent bearer services in data, as well as facsimile transmission. These services have been widely implemented and tested in public operation. As the trend in fixed data and telecommunication networks is towards higher user rates, also GSM has to evolve so that it fulfils new user requirements.

A set of GSM enhancements specified by European Telecommunications Standards Institute (ETSI) is called GSM phase 2+. One of the main features in phase 2+ is High Speed Circuit Switched Data (HSCSD). The standardisation work started during 1994 and HSCSD belongs to the 1996 release of GSM phase 2+ standards. HSCSD is an enhancement of the current circuit switched GSM data services to cover higher user rates than 9.6 kbps. The architecture of the HSCSD service is based on the physical layer of the current data services. The increased user rate is achieved by using a combination of more than one full rate traffic channel (TCH/F) for a single connection.

In the transparent data transmission mode the maximum HSCSD data rate is 64 kbps using bit transparent protocols. In non-transparent mode the maximum data rate is 38.4 kbps achieved by using a combination of four TCH/F9.6 channels. With the new 14.4 kbps data channel, the maximum non-transparent data rate is increased to 57.6 kbps. The effective user data rate of the non-transparent HSCSD data service may be further increased by using GSM data compression that is based on V.42bis algorithm. This compression algorithm increases the effective user data rate to between two and four times the physical user data rate.

The main applications for HSCSD are, for example, facsimile, transfer of large messages, WWW browsing, and video in cellular environment. Most of the facsimile equipment operate with data rate 14.4 kbps, and rates up to 28.8 - 33.6 kbps will be used in the future. Transmission of large messages, including electronic mail and file transfer, benefits significantly from the introduction of HSCSD since the transmission time can be decreased. The usage of WWW (and other graphical, user friendly applications) with HSCSD is more convenient, since the interactivity can be improved with faster transmission of images. However, from the network utilisation point of view HSCSD is not optimum for WWW, since for most of the connection time the user is reading the transferred pages. Audio-visual communication (e.g., ITU-T H.324 protocol family for audio-visual communication) is an interesting field of service for HSCSD. The development in low bit rate video coding algorithms (e.g., ITU-T H.263) enable video retrieval and conferencing services with the data rates achievable using HSCSD. There is huge potential for future applications in this area.

2. HSCSD STANDARD

High investments have been made in the existing GSM network equipment, and thus the development of the HSCSD-feature follows the existing architecture as much as possible. The physical characteristics have been kept in the air interface, as well as in the network infrastructure. Majority of the extensions are included into the data and signalling protocols.

In the call establishment phase the user negotiates a set of HSCSD parameters, e.g., user rates, number of channels, modem, possible channel codings, with the network. These parameters can be entered to the mobile equipment using the GSM AT commands. Software manufacturers can implement PC applications using AT commands and therefore the user does not have to study the specified commands. Alternatively, the mobile equipment can have special user interface designed for parameter setting. However, most of the users will rely on the default values given by the manufacturers at the implementation time.

An HSCSD connection is established if the network can provide a service within the limits of the given parameters. The connection is kept open until the user gives a termination command, provided that the service characteristics remain within the given limits throughout the connection.

Charging of an HSCSD call is based on the usage of TCH/F channels, as in the current GSM system. The calling HSCSD mobile will pay for all the used channels, but the called party may have to pay for the additional (exceeding one) TCH/F channels when the call is originated from a fixed network. The number of channels and the channel coding used may be included in the charging principles. [2]

2.1 System Architecture

An HSCSD configuration consists of multiple independent full rate traffic channels (TCH/F) between Mobile Station (MS) and the network, as shown in Figure 1. The channel types that are allowed in HSCSD are TCH/F4.8, TCH/F9.6, and TCH/F14.4, which provide data rates 4.8 kbps, 9.6 kbps, and 14.4 kbps per channel, respectively. The standard for 14.4 kbps data channel is written so that the TCH/F14.4 is an applicable channel type for HSCSD configurations. Operation of each individual TCH/F14.4 as part of an HSCSD-connection is kept as it is with TCH/F9.6 and TCH/F4.8.

For cost optimisation reasons, the operation of Base Transceiver Station (BTS) is modified as little as possible. In the network architecture, most of the modifications concentrate on the Base Station Controller (BSC) and Mobile Switching Center (MSC). In an HSCSD configuration, the data stream is split into sub-streams between split/combine-functions. One split/combine-function is located in the mobile station and the other either at the MSC's Interworking Functions (IWF) or the Base Station Subsystem (BSS).

The location of the split/combine-function in the network side depends on the used channel configuration; in configurations using up to four TCH/F channels, the split/combine functionality is located at the IWF. Between BSS and IWF the combination of individual channels are carried within one 64 kbps link. When using more than four channels, the split/combine functionality is taken care of by the BSS. Then the overall data stream between BSS and IWF is carried in a 64 kbps physical ISDN link.

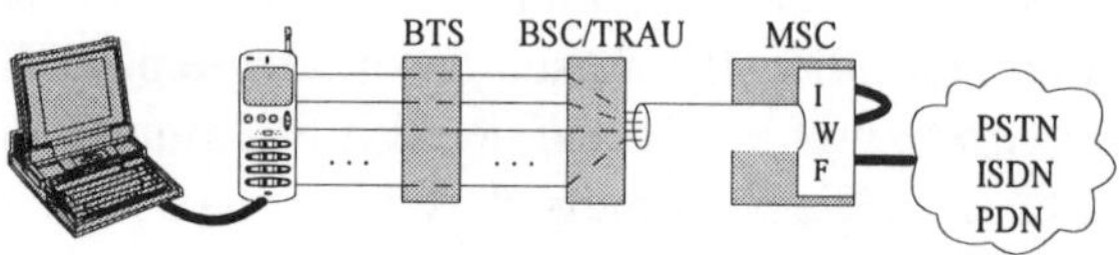

Figure 1. HSCSD system architecture with multiple channel configuration

2.2 Air Interface

In the GSM air interface one 200 kHz carrier is divided into 8 TDMA time slots, one of which is used for each TCH/F. In an HSCSD configuration, multiple TCH/F channels can be allocated for one mobile station. The main target has been re-usability of the existing implementations. Channel coding and interleaving of the TCH/F channels are kept unmodified, as are the rate adaptations between the split/combine functions. All channels in the HSCSD configuration follow the same frequency hopping sequence. Choosing the same hopping sequence keeps the radio synthesiser from switching the frequency between consecutive reception/transmission time slots within a TDMA frame. The same training sequence is used for each channel. In the encryption, an enhancement to the existing protocols has been standardised to achieve improved security. A separate ciphering key is generated for each of the channels. The keys are generated from the one ciphering key derived in the call establishment.

Allocation of several channels for data transfer increases the complexity of mobile station. All the existing mobile stations are designed for operation where the radio reception and transmission never overlap, i.e., half duplex radio operation. For symmetric transmission of up to 2 TCH/F channels, the overlapping can be still avoided. However, for higher data rates the radio frequency architecture must be re-designed, since the reception and transmission start to overlap.

HSCSD includes a possibility for asymmetric operation. This allows reception with higher data rate than that used for transmission while still allowing half duplex operation of the radio block in mobile station. Examples of asymmetric configurations are 2+1, 3+1, 3+2, and 4+1; where the first number indicates the number of downlink TCH/F channels and the last number the number of uplink TCH/F channels. In addition to those hardware requirements brought about by the overlapping of reception and transmission, the capability of the radio synthesiser for quick switching between transmission and reception, as well as processing requirements, are important design issues.

The capability of the mobile equipment is indicated by a mobile station multislot class parameter. There exists 18 multislot classes that were defined while the HSCSD was specified. The multislot class indicates the configurations that can be supported by the mobile equipment, i.e., the maximum number of channels supported in symmetric or asymmetric mode of operation. The class is indicated to the network at call establishment, to enable the network to select a suitable configuration based on the capabilities of the mobile equipment and the traffic situation in the network at the very moment.

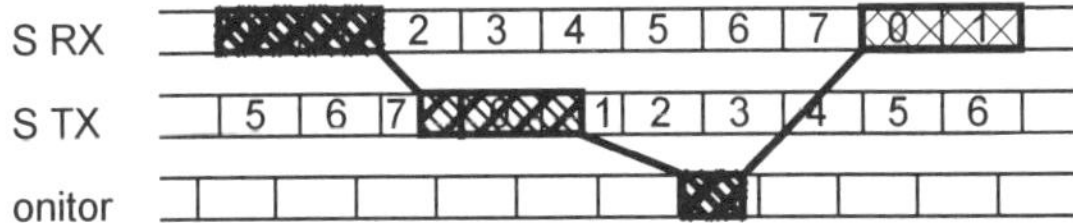

Figure 2. An example for the symmetric HSCSD configuration with two TCH/F channels

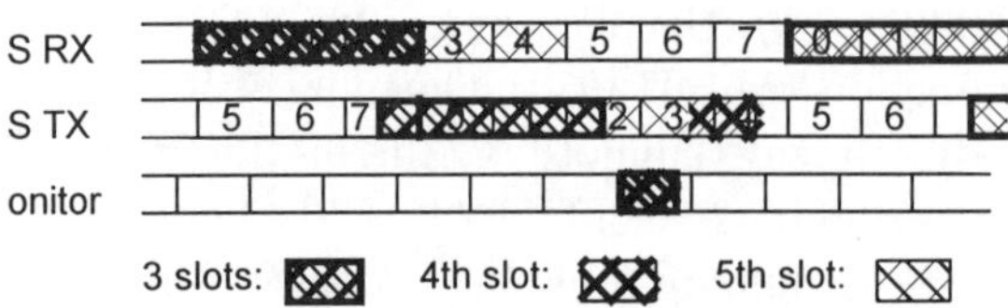

Figure 3. An example for the symmetric HSCSD configuration with more than two TCH/F channels. Overlapping of the reception and transmission can be seen in the figure

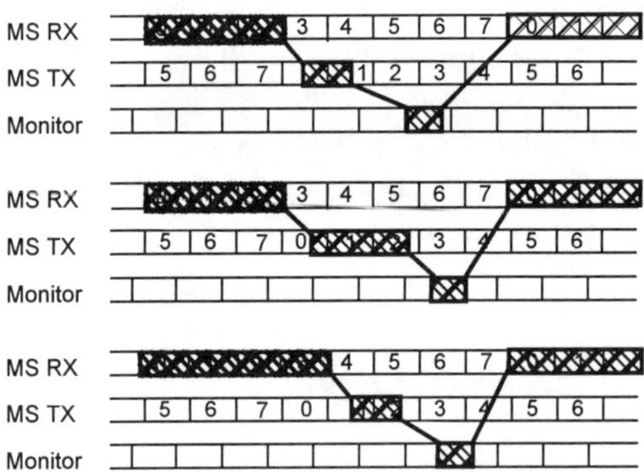

Figure 4. An example for the asymmetric HSCSD configurations for 3+1, 3+2, and 4+1 TCH/F channel configurations

For controlling the operation of a TCH/F channel, a Slow Associated Dedicated Control Channel (SACCH) and a Fast Associated Control Channel (FACCH) are used. SACCH is used mainly for reporting the results of neighbouring cell monitoring and reception signal level and quality to the network. FACCH is used when higher signalling rate is needed, e.g., during a handover.

In symmetric HSCSD configuration, an individual SACCH is associated with each of the channels. This enables separate power control for each channel. Also in asymmetric HSCSD configurations where also uni-directional channels are part of the connection, each of the bi-directional channels report the strength and quality of the received signal separately, enabling separate power control for each of the bi-directional channels. For the uni-directional channels, an SACCH is associated with the downlink direction only. The worst channel quality and signal strength among the uni-directional channels and the bi-directional main channel is reported on the main channel of the HSCSD configuration. The main channel is the only channel in an HSCSD configuration carrying FACCH information. The beginning of the SACCH frames in each of the channels is synchronised to the start time of the SACCH frames in time slot number zero. This assures that the idle slot in the 26 TDMA frame multiframe is during the same TDMA frame for all the channels. Thus the idle TDMA frame can be used for achieving synchronisation with neighbouring cells.

2.3 Data Protocols

Both transparent and non-transparent bearer services are supported in HSCSD. Most of the functionality of the data protocols is located at mobile station's Terminal Adaptation Function (TAF) and IWF. However, when more than four channels are used, the split and combine function is located at the Base Station Subsystem (BSS).

2.3.1 Transparent Data

Transparent data service is based on the V.110 protocol that has been specified for ISDN. The required rate adaptation between 16 kbps ISDN signalling rate and 12 kbps air interface rate is performed by leaving 20 bits, e.g., mainly synchronisation bits, out of the 80-bit V.110 'intermediate rate' frame.

The challenge in HSCSD is to retain the order of information bits in data transfer when the parallel datastreams are combined. Due to the multiple channel configurations in HSCSD operation, in the V.110-frames there are redundant status bits[1] that can be used for synchronisation

[1] The status of the V.24 circuits is transmitted over GSM in status bits; SA (4 times), SB (3 times), and X (twice). In the multiple channel HSCSD configuration, it is possible to use the redundant status bits (two SA, one SB, and one X) for the numbering up to eight channels without reducing the rate of the status bits per connection. One extra bit can be used for frame synchronisation between channels in case of time dispersion.

between channels. Three status bits per V.110-frame are used for channel numbering and one bit carrying a 31-bit pseudo-random sequence is used for synchronisation between the channels. The 31-bit pseudo-random sequence allows fast synchronisation in case the synchronisation between the channels has been lost.

Rate adaptation for the data rates that are not divisible by the air interface rate of the HSCSD channels can be performed by padding in the channel carrying the highest substream number. For example, 14.4 kbps user rate can be achieved with two TCH/F9.6 channels when only half of the 'user' data bits are used in the other channel. Rest of the bits are padded with fill bits.

2.3.2 Non-transparent Data

Non-transparent data service is based on a new version of the Radio Link Protocol (RLP) between mobile station and IWF. The RLP takes care of the frame numbering and re-transmission of corrupted frames. This enables delivery of the correctly received frames in the order of transmission.

The version 2 of the RLP has a frame header lengthened with additional eight bits;the send and receive frame numbering fields are both extended with three bits to enable the use of extended frame numbering space. An enhancement has been introduced also to the Layer 2 Relay (L2R) functionality. In RLP version 0 and 1, minimum of one L2R status octet is transmitted in each of the RLP frames. However, in RLP version 2 the L2R status octet is transmitted only if there is need for it. Therefore the theoretical maximum data rate per channel is kept, even if there is one additional octet in the RLP header.

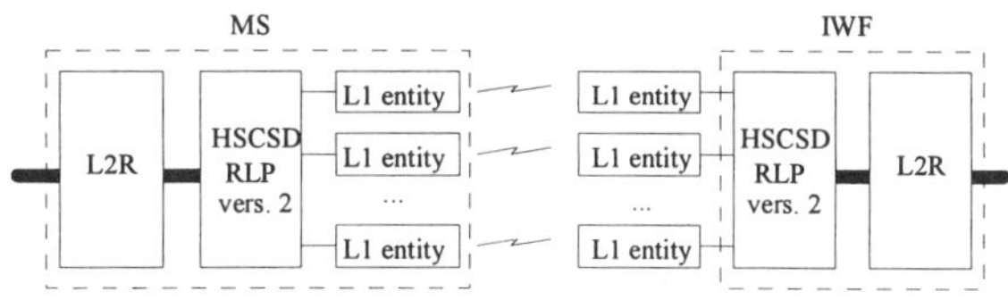

Figure 5. Protocols for non-transparent HSCSD; Radio Link Protocol combining and splitting multiple channels

2.4 Signalling

The current GSM system has separate bearer services specified for all the user rates. However, HSCSD configuration uses General Bearer Services and the user rate is just one of the parameters of the bearer service. The user rate can be negotiated at the call establishment and even during the connection in case of the non-transparent data operation.

In the call establishment phase, a set of HSCSD parameters is negotiated between the user and the network. During the signalling, the mobile station transmits its Classmark information (including multislot capability of the mobile station) to the network. The HSCSD configuration related parameters are included in the bearer capability information element of the Setup message. The parameters included at the bearer capability information element are:

- Maximum number of TCH/F channels to be used (Max TCH/F)
- Acceptable channel codings in the air interface (ACC)
- Other modem type (OMT, e.g., V.17, V.32bis, or V.34)
- Fixed network user rate indicating the data rate the GSM PLMN is using to interwork with external network (FNUR)
- Wanted air interface user rate (AIUR, only for non-transparent data)
- User initiated modification of channel configuration allowed or not (UIMI, only for non-transparent data)

In the mobile originated call establishment, the network informs the mobile station about the selected configuration with Call proceeding message that includes the OMT, FNUR, and UIMI parameters. The UIMI is used only for non-transparent data. Each of the radio channels are activated separately. A description of the multislot configuration is given to the mobile station in the Assignment command message. The Assignment complete message acknowledges the given configuration and the IWF can allocate appropriate resources for the call.

In the mobile terminated call establishment, the Setup message includes OMT, FNUR, and UIMI parameters. UIMI is used only for non-transparent data. The mobile station acknowledges with Call confirmed, including additionally the ACC, Max TCH/F, and AIUR parameters. AIUR is used for non-transparent data only. The procedure is followed by the activation of the channels and the assignment procedure, as in the mobile originated case.

During an HSCSD call, the mobile station may move between cells, in which case a need for a handover appears. The mobile station measures the signal levels in the serving and the neighbouring cells, as in the current GSM system. Measurement reports are delivered to the network and in case the call can be served better in the neighbouring cell, the handover for all the channels of the HSCSD connection is initiated. In the intra BSC handover, all the parameters of the HSCSD call are known by the BSC. The BSC activates channels in the new cell, and the description of multislot configuration is given to the mobile station in the Handover command message. The handover procedure is completed with the Handover complete message, after which the IWF can adjust appropriate resources. Simultaneously the radio channels in the old cell are released.

In inter BSC handover, the new BSC must be informed about the parameters negotiated at the call establishment. The serving MSC forwards required parameters to the new BSC and the handover procedure continues as in the intra BSC handover. In an inter MSC handover there is need for parameter exchange between the MSCs, to enable delivery of parameters to the new BSC. The messaging between MSCs is carried using the Mobile Application Part (MAP) signalling messages.

During the call, it is possible that the network conditions (e.g., network load or channel quality) vary. Therefore, the network may want to increase or decrease the number of channels allocated for an HSCSD configuration. In case there is lack of channels in the cell, the number of TCH/F channels in some of the configurations may be temporarily decreased. This can be done with a new Configuration change procedure. The description of the new multislot configuration is sent to the mobile station, and based on this acknowledgement the IWF may adjust it resources. The Configuration change procedure is possible only within the limits defined by the parameters at call establishment. In non-transparent data, the number of TCH/F channels can be varied since the user rate may be variable. However, in transparent data, the user rate is fixed and the Configuration change is possible only if the used channel coding is changed simultaneously.

In case the user is not satisfied with the parameters negotiated at the beginning of the call, the parameters can be changed during the call. The user may want to increase or decrease the user rate if the requirements for communication change. A Modify command is then sent by the mobile station to the MSC, after which the network may upgrade or downgrade the number of radio channels as requested.

ACKNOWLEDGEMENTS

The author gratefully acknowledge the colleagues from Nokia for a great effort in research and standardisation of HSCSD. The contribution of the Nokia team to the GSM HSCSD standard has been significant.

REFERENCES

European Telecommunications Standards Institute, GSM 02.34, High Speed Circuit Switched Data (HSCSD) - Stage 1.

European Telecommunications Standards Institute, GSM 03.34, High Speed Circuit Switched Data (HSCSD) - Stage 2.

European Telecommunications Standards Institute, SMG#21, The set of Change Requests for HSCSD, February 1997.

J.Nieweglowski and T. Leskinen: 'Video in mobile networks', Proc. of the European Conference on Multimedia Applications, Services and Techniques ECMAST 96, May 1996.

J. Hämäläinen: Design of GSM High Speed Data Services, Doctor of Technology Thesis, Tampere University of Technology Publications 186, October 1996.

Chapter 5

CAMEL and Optimal Routing

Evolution of GSM Mobility Management towards UMTS

MARKKU VERKAMA
Nokia Telecommunications

Key words: CAMEL, Intelligent Network, Mobile Application Part, Mobility Management, Operator Specific Services, Routing.

Abstract: Mobility management is an essential functionality in mobile networks enabling subscribers to roam outside their home network. Two new features will be introduced to the GSM system that are related to mobility management. The other feature called CAMEL supports provision of operator specific services, while the other one enables optimal routing of connections. This chapter describes the objectives and technical realization of CAMEL and optimal routing. The implementation of these features makes GSM Mobile Application Part a full fledged mobility management platform capable of meeting the challenges of third generation mobile networks.

1. INTRODUCTION

A unique feature of mobile systems such as GSM is that mobile users who subscribe to one operator's services are able to seamlessly roam in other operators' networks, provided that necessary agreements have been made by the operators and the networks support the same services. The functions needed to provide this feature are part of mobility management. The cornerstone of mobility management in the GSM family systems (GSM 900, GSM 1800, GSM 1900) is the GSM Mobile Application Part (MAP) protocol (GSM 09.02). MAP is used in operations such as authentication of subscribers, transfer of subscriber data from the home network to the serving network, and routing of calls towards the mobile subscriber.

GSM phase 2+ standardization includes two exciting work items that are closely related to mobility management. These work items are "Customised Applications for Mobile network Enhanced Logic" (CAMEL) and "Support of Optimal Routing." The purpose of CAMEL is to enable provision of operator specific services in GSM networks, while the other aims at optimizing use of circuits in certain call scenarios. Both work items add important new features to GSM MAP and pave its way towards a general mobility management platform that meets the requirements of third generation mobile systems (see Verkama et al, 1996).

The purpose of this chapter is to describe briefly these new GSM features, including the main principles of the technical solutions for realizing them. The presentation is mainly based on the so called Stage 1 and Stage 2 specifications of CAMEL and optimal routing (GSM 02.78; GSM 02.79; GSM 03.78; GSM 03.79). Stage 2 specifications GSM 03.78 and 03.79, which describe technical realization, were approved by the Special Mobile Group Technical Committee (SMG TC) of ETSI in April 1997.

Both CAMEL and optimal routing specifications concern only the network, not terminals, and therefore thc scope of the chapter is strictly in network issues. Nevertheless, the introduction of operator specific services by CAMEL also poses new challenges for mobile terminals, which have to provide user interfaces hiding the complexities of different services (Ali-Vehmas, 1997).

The remainder of the chapter is organized as follows. Section 2 provides a brief introduction to the GSM MAP protocol and discusses its underlying service provisioning philosophy. The objectives and technical realization of CAMEL are then presented in section 3. Section 4 deals with optimal routing, and finally some open issues are discussed in section 5.

2. GSM MOBILE APPLICATION PART

MAP is used in several interfaces of the GSM system, as is illustrated in *Figure 1*. The network elements concerned in GSM phase 2 are mobile switching center (MSC), home location register (HLR), visitor location register (VLR), and equipment identity register (EIR). MAP uses the services of TCAP in SS7 signalling network, except in the B interface, which is internal to MSC and its associated VLR. In the open interfaces, C,

D, E, F, and G, MAP is currently used for the following main mobility management procedures:

- MSC- HLR interface (C): call routing
- HLR-VLR interface (D): location management, management of subscriber data and security information, call routing
- MSC-MSC interface (E): inter-MSC handover
- MSC-EIR interface (F): international mobile equipment checking
- VLR-VLR interface (G): transfer of security information

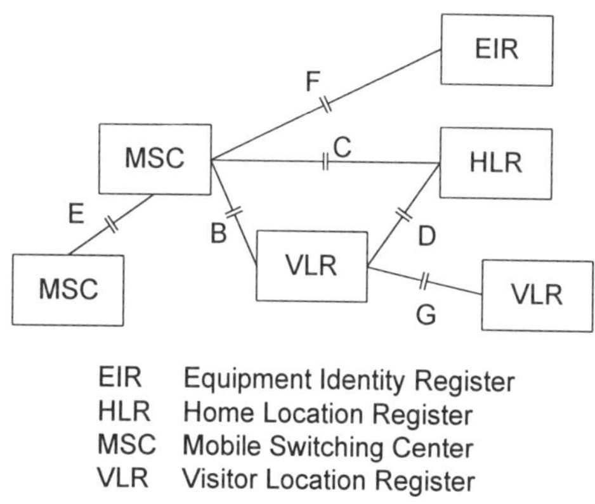

Figure 1. GSM network interfaces where MAP applies (phase 2).

Additionally, MAP is used for handling supplementary services, routing short messages, and for operation and maintenance services. A completely new MAP interface is being specified for the General Packet Radio Service (GPRS) in phase 2+ standardization (cf. chapter 2.3 in this volume). This interface is between serving GPRS support node and HLR (or GPRS register) and is used for example to transfer subscriber data and authentication triplets. Notice that in GSM architecture the authentication center that manages security information is a functional unit of the HLR.

When a subscriber roams outside her home network, the visited network should ideally provide the same set of services as the home network. This means that information must be exchanged between the home network and the visited network. In GSM basically all service information is transferred by MAP from the home location register to a visitor location register in the serving network.

To understand the requirements of roaming and the nature of subscriber data, it is useful to divide services into two categories:

1. Basic communication services such as speech service, mobile terminated short message service, or 9600 bit/s asynchronous data service. This kind of services are commonly known as bearer services and teleservices, and the related subscriber data consists of codes identifying the services that have been subscribed to.
2. Supplementary services that modify the basic services, often through some logic. For these services, subscriber data typically consists of subscriber specific parameters pertaining to the service logic, such as the forwarded-to number to be used when a call is forwarded because of no reply.

One can say that the services in the first category typically require hardware support in the serving network, such as modems or interworking units. The capabilities or limitations of these will apply equally to all subscribers accessing the network, whether roaming or in their home PLMN, and regardless of the mobility management approach.

The services in the second category are more software oriented. In order for a roaming subscriber to have access to a supplementary service, the visited network must have access to the subscriber specific service parameters and the service logic. It is here where the traditional MAP approach of transferring only parameters can be limiting, because the visited network then should implement the service logic. Certain supplementary services related to terminating calls, such as unconditional call forwarding, are an exception as they can be handled entirely by the home location register.

Seamless service across different GSM networks requires thus rigorous standardization and basically implementing the same services in all networks. The advantage is that this allows a unified service environment across different networks and facilitates a common approach to solving service conflicts, when new services are introduced. On the other hand, a drawback is that operators cannot independently provide new services unless service provisioning is strictly limited to the home network. This is why the new CAMEL feature is being introduced to GSM.

3. CAMEL

3.1 Purpose

An important way for operators to differentiate themselves in an increasingly competitive environment is to provide innovative services. Hence the development and deployment of new services should be fast. This desire is not in line with the requirements of standardization and ubiquity in the mobile network environment, where roaming between different networks is essential.

The intelligent network (IN) concept was developed to address these needs in a fixed (wireline) network environment (Q.1201). The intelligent network distinguishes between service switching, service control, and service data functions. IN can be used in today's GSM networks; however, it is restricted to service provisioning within the home network because it lacks mechanisms to support the mobile multioperator environment of GSM. This is what CAMEL addresses.

The purpose of CAMEL is to provide mechanisms for supporting services consistently and independently of the subscriber location. CAMEL extends the intelligent network concept by enabling service control of operator specific services outside of the serving GSM network. This is achieved by arranging new information exchange mechanisms between the serving network, home network, and so called CAMEL Service Environment. The CAMEL Service Environment is a logical entity that processes activities pertaining to operator specific services.

In GSM 02.78 operator specific services are defined to mean any services offered in a GSM network but not standardized by the GSM specifications. An example could be a number translation service where a code like '123' gives access to the same service independent of the serving network. Another example could be a call forwarding service where the forwarded-to number depends on the location of the called subscriber or the time of day or both.

CAMEL procedures can be applied to all circuit switched basic GSM services (except emergency calls). Non-call related events such as call independent supplementary services procedures, transfer of short messages, or mobility management procedures are not in the scope of CAMEL.

From the perspective of Universal Mobile Telecommunications System (UMTS), CAMEL is important as a step toward the virtual home environment, which is one of the UMTS requirements (UMTS 22.01). The virtual home environment means roughly that subscribers should get the same set of services, features, and tools independent of their location.

3.2 Principles

CAMEL is a network feature and its implementation involves many network elements. Two main components can be identified in the solution:

- use of the intelligent network
- use of GSM MAP to transfer CAMEL information between appropriate network elements

Let us first concentrate on the intelligent network part. Using the intelligent network concept means specifying basic call state models, service control function, service switching function, and their interface for GSM. These concepts have been modified for CAMEL from the corresponding ITU-T and ETSI specifications for fixed (wireline) networks.

According to GSM 03.78, a basic call state model "provides a high-level model of GMSC- or MSC/VLR-activities required to establish and maintain communication paths for users." Here GMSC refers to the gateway MSC involved in mobile-terminated calls.

To be more specific, a basic call state model describes the call related processing by listing so called detection points (DP) and points in call, and by identifying possible transitions between these. A *point in call* identifies MSC/VLR or GMSC activities related to call states that may be of interest to service logic instances. *Detection points* are stages in call processing at which notifications or control information can be exchanged between the switching entity and the service logic. Detection points cause actions only if they are armed, and there are three types of them in CAMEL. *Triggers* are statically armed detection points that always cause call processing to be suspended until further instructions. *Request events* cause call processing to be suspended as well but are armed dynamically by the service logic. *Notify events*, which can also be armed dynamically, cause only a notification to be sent without suspension of call processing.

Service control function is the functional entity in the intelligent network that contains the service logic for implementing a particular service. *Service switching function* is the functional entity that interfaces the switch to the service control function. In CAMEL these are called

GSM Service Control Function (gsmSCF) and GSM Service Switching Function (gsmSSF), respectively. The network element where service control function resides is usually called service control point.

The protocol defined for the gsmSSF-gsmSCF interface is called CAMEL Application Part (CAP) and is specified in GSM 09.78. Its counterpart in fixed (wireline) networks is called Intelligent Network Application Protocol (INAP). INAP is being specified in, so called, capability sets, and CAP is based on a subset of ETSI's capability set 1 INAP (ETS 300 374-1). Because call processing in GSM differs from fixed (wireline) networks, some parts of INAP are irrelevant to GSM and are thus not needed in CAP. On the other hand, CAP has some mobile specific additions. Like its wireline counterparts, CAP uses TCAP services in the SS7 signalling network.

When a GSM subscription includes operator specific services, two cases can be identified: services related to mobile originated and mobile terminated calls. In both cases CAMEL subscription information must be transferred from the home network to the GSM Service Switching Function in order to initiate the information exchange with the GSM Service Control Function. The solution in CAMEL is to use GSM MAP to transfer the information from the home location register to the serving MSC/VLR or GMSC, as will be explained in more detail below.

The key elements of the CAMEL feature are MAP and CAP protocols, home location register, and the new functional entities gsmSSF and gsmSCF (see *Figure 2*). The functional architecture shows also an interface between HLR and gsmSCF. This MAP based interface enables so called any time interrogation, where gsmSCF can interrogate home location register for the state and location information of any subscriber at any time.

CAMEL is being specified in phases. The specifications which this chapter is based on define phase 1 of the CAMEL feature. In addition to mobile originated and mobile terminated calls and any time interrogation, CAMEL supports forwarded calls, which are treated like mobile originated calls, and suppression of announcements.

The following subsections provide a more detailed description of the CAMEL feature in the context of mobile originated and mobile terminated calls.

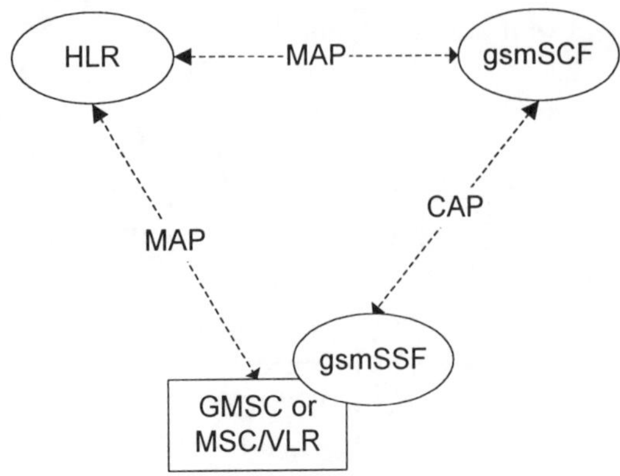

Figure 2. Key elements of the CAMEL feature; GMSC is involved in the terminating CAMEL and MSC/VLR in the originating CAMEL.

3.3 Mobile Originating Calls

From service perspective normal mobile originating calls involve only the serving MSC/VLR. When CAMEL is applied, gsmSSF and gsmSCF are involved as well. In this case, gsmSSF is a functional entity in the serving switch. The basic call state model of a mobile originating call is shown in *Figure 3*.

Table 1. Detection points for originating calls

Detection Point	Type	Description
DP2 Collected_Info	Trigger	Originating CAMEL subscription information has been analyzed.
DP7 O_Answer	Event - notify	Call has been accepted and answered by the terminating party.
DP9 O_Disconnect	Event - notify Event - request	Disconnect indication has been received from either one of the parties.

The main functions associated with the points in call have been listed in the figure. Their exact definition can be found in GSM 03.78. Detection points are described in *Table 1*.

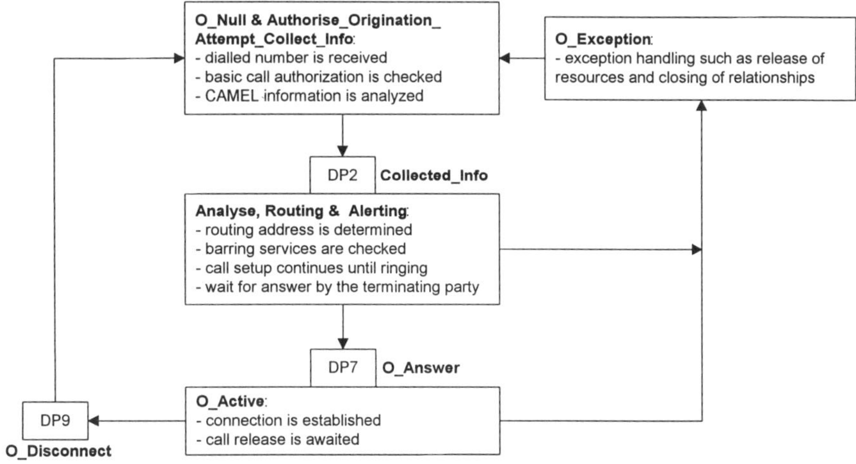

Figure 3. CAMEL basic call state model for mobile originating calls; the main functions of the points in call are indicated (GSM 03.78).

If the user's subscription includes operator specific services that apply to originating calls, an information element called "originating CAMEL subscription information" (O-CSI) is included in the subscriber data in home location register. This information element contains:

- gsmSCF address (E.164 number)
- service key that identifies the service logic to be applied by gsmSCF
- information about default call handling in case of error in the dialogue between gsmSSF and gsmSCF
- list of triggering detection points (currently only DP2 "Collected_Info" for originating calls)

When the subscriber registers in the serving network, the originating CAMEL subscription information is copied to the visitor location register along with other subscriber data.

The message flow related to an originating call is illustrated in *Figure 4*. Upon receiving the call setup message, the MSC sends a "send info for outgoing call" (SIFOC) message to the visitor location register (over an internal interface). After performing some basic checks (e.g. call barring) the VLR responds with a "complete call" message that includes the originating CAMEL subscription information. The message informs the MSC that CAMEL should be applied. This means encountering DP2 in the basic call state model, which triggers the gsmSSF in MSC to initiate a dialogue with gsmSCF by sending an InitialDP message. For an originating call, the mandatory information elements in this message include IMSI, called party number, location information, and the service key. The location information includes cell or location area identity, VLR address, the age of the location information as well as geographical location information if available.

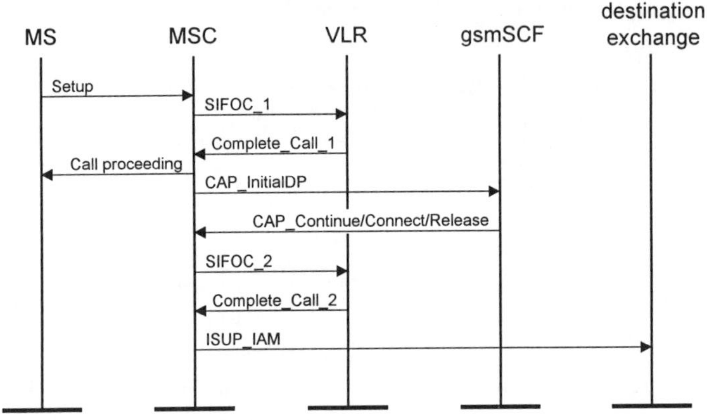

Figure 4. High level message flow for CAMEL originating call; gsmSSF is in MSC.

The response from the gsmSCF is one of the following:

- *Continue*, in which case call processing continues without modifications
- *Connect*, in which case call processing continues with modified information such as a new destination address,
- *Release call*, in which case the call shall be released.

If the response is either Continue or Connect, another "send info for outgoing call" message is sent to the VLR where further checks are performed. This is necessary because call parameters may have changed. Having received the final "complete call" response, the MSC can finally start call setup towards the destination exchange.

Once a gsmSSF-gsmSCF dialogue has been opened, gsmSCF may arm other detection points dynamically by sending a "request report basic call state model event" message. This message indicates which event detection points shall be armed and how, i.e., whether they will be request events or notify events. After the call is accepted and answered, the other detection points DP7 and DP9 are then encountered, and they may or may not be armed depending on the needs of the particular operator specific service.

3.4 Mobile Terminated Calls

For mobile terminated calls, the CAMEL feature is quite complex due to routing interrogations and call forwarding scenarios. The terminating call state model describes the actions of the gateway MSC (GMSC) that interrogates the home location register for routing information. The model is shown in *Figure 5*. The main functions of the points in call are indicated in the figure, and exact definitions can again be found in GSM 03.78.

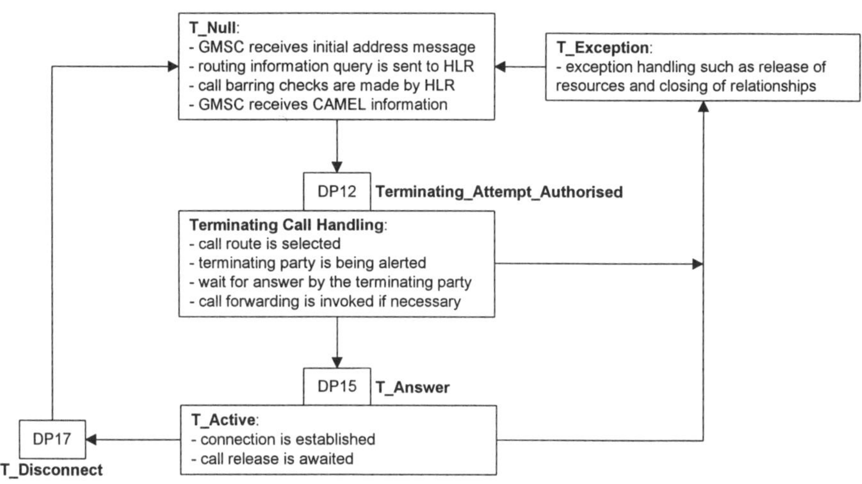

Figure 5. CAMEL basic call state model for terminating calls; the main functions of the points in call are indicated (GSM 03.78).

An information element called "terminating CAMEL subscription information" (T-CSI) is included in the subscriber data if the user's subscription includes operator specific services that apply to terminating calls. This information element contains the same parameters as the corresponding information element for originating calls, O-CSI, but the calls. This information element contains the same parameters as the

Table 2. Detection points for terminating calls

Detection Point	Type	Description
DP12 Terminating_Attempt_ Authorised	Trigger - request	Terminating CAMEL subscription information has been analyzed.
DP15 T_Answer	Event - notify	Call has been accepted and answered the terminating party.
DP17 T_Disconnect	Event - notify Event - request	Disconnect indication has been receiv from either one of the parties.

corresponding information element for originating calls, O-CSI, but the parameters may have different values. The only allowed trigger point for terminating calls is currently DP12 "Terminating_Attempt_Authorised."

For simplicity, consider a situation where neither CAMEL nor call forwarding by the home location register modify the call destination, as illustrated in *Figure 6*. After receiving the Initial Address message (IAM) from the originating exchange, GMSC sends a "send routing information" (SRI) query to the HLR. The query must indicate the CAMEL phases supported by the GMSC. If the subscriber has a terminating CAMEL service active, the HLR returns an acknowledgement to the GMSC with the terminating CAMEL information, instead of initiating a roaming number query. This indicates to the GMSC that CAMEL should be applied.

Optionally, HLR may also initiate a "provide subscriber information" (PSI) procedure towards the visitor location register serving the subscriber, in which case location information and subscriber state of the called party are included in the "send routing information" acknowledgement. GSM 03.78 is vague about when the status and location information query is to be performed, but at least one situation is related to optimal routing as described later in section 4.2.

Having received the terminating CAMEL information from the home location register, the GMSC suspends further call processing and initiates a dialogue with gsmSCF by sending an InitialDP message. Location information and subscriber state are indicated if they have been received from the home location register. The response from the gsmSCF may again be Continue, Connect, or Release.

If the response is Connect, meaning that some call information has been modified, the GMSC sends an Address Complete message (ACM) towards the originating exchange. The purpose is to stop any call timers so that no problems would occur due to delayed call processing. In addition to possible modifications in call parameters, the Connect message may indicate that any announcements and tones due to unsuccessful call setup should be suppressed in the serving MSC.

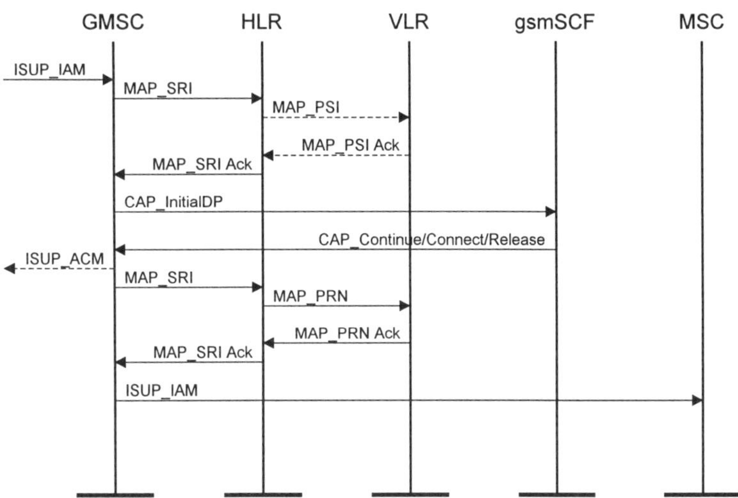

Figure 6. Message flow for a terminating call; no call forwarding; gsmSSF is in GMSC.

If the gsmSCF response is either Continue or Connect but does not modify the destination number, another routing information query is sent to the home location register. This message shall contain a CAMEL suppress information element to indicate that the CAMEL dialogue is already in progress. This then leads to a "provide roaming number" (PRN) query from the visitor location register and the roaming number being relayed back to the GMSC.

When the GMSC receives the roaming number it continues call setup towards the serving MSC. Again, other detection points may cause message exchange between the gsmSSF and gsmSCF during the call processing if they have been armed by the gsmSCF.

Several alternative scenarios are also covered by the specifications, such as

- call is forwarded by the home location register; originating CAMEL may apply to the forwarded call
- call is forwarded by an operator specific service in the gsmSCF; originating CAMEL may apply to the forwarded call

- interaction of any call barring with the above
- effects of closed user group

It can also be noted that GMSC, which performs routing information interrogation for mobile terminated calls, is always in the home PLMN when traditional call routing applies. However, optimal routing can change that for mobile-to-mobile calls as will be explained in the next section.

4. OPTIMAL ROUTING

4.1 Motivation and Principles

One reason for the success of GSM has been its international roaming capabilities. Considering the charges incurred to the call parties, the principle is simple and clear. A call to a GSM subscriber is always routed towards her home PLMN based on the dialled number, MSISDN. This call route is called the *home PLMN route*, and the calling party pays only for the call leg to the home PLMN. The called party pays for any further legs from her home PLMN. This means paying international tariffs if the called party is roaming in another country. At least in Europe, the called party typically does not have to pay for air time, so charges from receiving calls incur often only due to international roaming. National roaming between GSM 900 and GSM 1800 systems may be an exception to this rule.

In certain international roaming scenarios this principle can lead to charges that may seem unreasonable. For example, if Hans, a subscriber roaming in Germany, calls Mikko, a Finnish subscriber roaming also in Germany, the call is always routed via Mikko's home PLMN in Finland. In other words, both Hans and Mikko end up paying for international call legs even though the call parties are actually within the same country. In fact, both could be even served by the same network.

The purpose of optimal routing is to reduce the number of unnecessary calls legs in scenarios such as above. GSM MoU has imposed two important constraints that affect the route selection:

1. No subscriber shall pay more for an optimally routed call than under the present routing scheme.
2. One call leg shall be paid entirely by one subscriber (at least in the first phase of optimal routing).

If the optimal route can not meet these requirements, standard routing shall be used. Because of these requirements and to avoid detailed charge calculations, optimal routing is initially limited to calls that stay within the same country and to calls where the final destination is in the same country as the dialled MSISDN. One call forwarding can take place but multiple call forwarding is not in the initial scope of optimal routing. These limitations mean that route selection can be done by comparing E.164 numbers, e.g., dialled number vs. VLR address.

GSM 02.79 describes ten scenarios where optimal routing can be applied. Here we shall concentrate on two important cases. First, optimal routing of basic mobile-to-mobile calls is described in section 4.2. The scenario discussed in section 4.3 is a call to a roaming subscriber where the serving MSC discovers that the call should be forwarded back to home country. An example of such a scenario is a call forwarded to a voice mail box when the subscriber is busy or does not answer.

The purpose is to describe the functionality needed in these two scenarios, not to give a complete presentation of network functionality required to support the whole optimal routing feature.

4.2 Basic Mobile-to-Mobile Calls

The solution for optimal routing of mobile-to-mobile calls is straightforward, because the existing GSM functionality essentially supports it already. Call forwarding, error situations, and negative outcomes are ignored in the following discussion. Several network elements are involved, including the serving mobile switch of the calling party (VMSC-A) and the gateway switch in that PLMN (GMSC-A), the home location register of the called party (HLR-B), as well as the serving visitor location register and switch of the called party (VLR-B and VMSC-B, respectively). We shall assume that HLR-B and GMSC-A are in different PLMNs to make the case more interesting.

The message flow is shown in *Figure 7*. Once the VMSC-A has received a call setup and performed necessary authorization checks, it constructs an Initial Address message using the dialled number, and sends it to the gateway. GMSC-A recognizes the dialled number as belonging to another GSM PLMN, and sends a request for routing information to HLR-B with an indication that it is an optimal routing inquiry. Notice that VMSC-A and GMSC-A may actually be the same network element if the required gateway functionality is implemented in the serving mobile switch.

Upon receiving the inquiry, HLR-B checks whether one of the optimal routing criteria is fulfilled; i.e., whether GMSC-A is in the same country as VMSC-B, or whether HLR-B is in the same country as VMSC-B. If the first criterion is fulfilled the call stays within the same country, while the second criterion handles the situation in which the called party roams in her home country. A third condition, whether GMSC-A and HLR-B are in the same PLMN, is already excluded by our assumption. If the result of the check is positive, HLR-B sends a request for a roaming number to VLR-B, which again indicates that it is for an optimally routed call.

Finally, the roaming number is relayed to the gateway, which constructs an Initial Address message using the roaming number. An exception to further GSM call procedures is that if the call is answered, the gateway shall insert the final destination address in the Answer message so that the VMSC-A can generate a correct charging record.

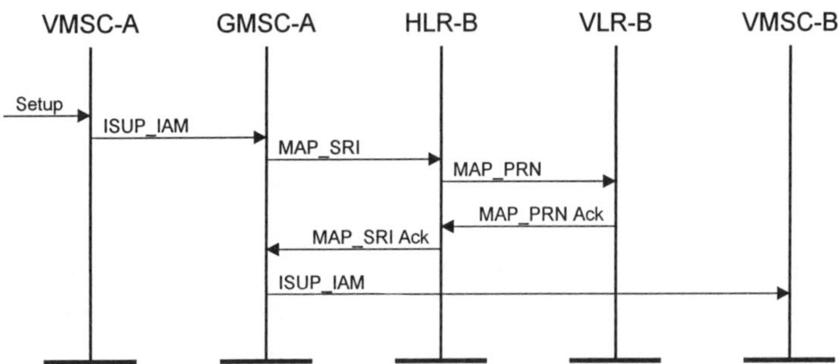

Figure 7. High level message flow for successfully optimally routed mobile-to-mobile call.

If the home location register determines that the charging requirements prevent optimal routing, it sends an inquiry for subscriber information to the VLR-B in order to determine whether the called party is detached from the network. Note that GSM mobile stations send a detach message when powering down, and there are also situations where the network can mark the user as detached. If the called party is not known to be detached, the home location register returns a negative response "optimal routing not allowed" to the GMSC-A, which then selects the home PLMN route for the call. If the subscriber turns out to be detached, the home location register will determine whether the "call forwarding on mobile subscriber not reachable" service should be applied.

In conclusion, functional differences between optimal routing and normal GSM routing are minor in this scenario. Essentially the gateway MSC has to recognize numbers belonging to other GSM PLMNs and the home location register has to check the optimal routing criteria. Nevertheless, the functional requirements of the whole optimal routing feature are more complex due to other scenarios, such as late call forwarding which is described below.

4.3 Late Call Forwarding

Late call forwarding means that the call is forwarded after having been extended to the serving network of the called party. An example is call forwarding on no reply. Late call forwarding, e.g. to a voice mail box, can incur double international call leg charges for roaming subscribers. Optimal routing will enable the release of unnecessary legs — if allowed by the charging requirements. The idea is that when the serving MSC detects that call forwarding should be applied, it requests the gateway MSC to resume call handling.

The following description assumes that the gateway MSC is in the same PLMN as the home location register (HLR-B); the message flow is shown in *Figure 8*. First the GMSC interrogates the home location register for routing information, which in turn sends a roaming number request to the visitor location register (VLR-B). The call reference number and GMSC address are relayed to the visitor location register in the request and stored there. Once the GMSC receives the roaming number, it constructs an Initial Address message towards the serving VMSC-B. When the VMSC-B receives the message and determines that the called party number belongs to its roaming number space, it sends a "send info for incoming call" (SIFIC) message to the visitor location register.

If the visitor location register, after appropriate procedures, determines that call forwarding should be applied, it returns an acknowledgement that indicates that the call is to be forwarded and that GMSC address is available. VMSC-B constructs then a "resume call handling" (RCH) request to the GMSC, which contains

- call reference number
- basic service group
- IMSI

- forwarded-to number
- forwarded-to subaddress if available
- forwarding reason (busy, no reply, not reachable)
- indication whether calling party is to be notified about forwarding
- possibly closed user group information
- possibly originating CAMEL subscription information (originating CAMEL applies to forwarded calls)

If optimal routing to the forwarded-to number is allowed from the GMSC, it acknowledges the resume call handling request, releases the circuit to the VMSC-B, and constructs an Initial Address message using the forwarded-to number.

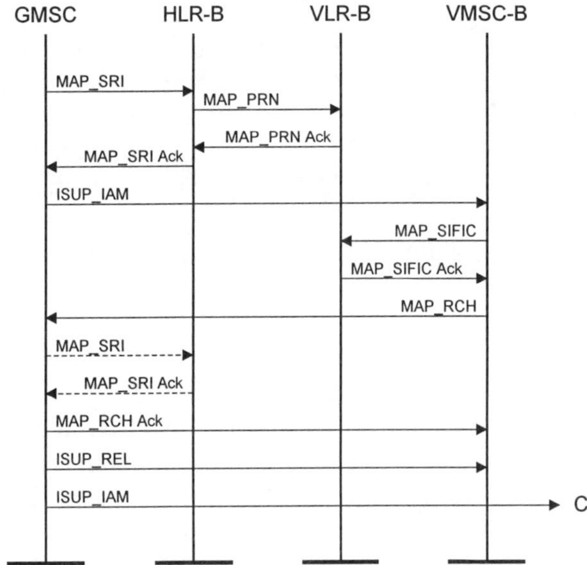

Figure 8. Message flow for optimally routed late call forwarding

An additional forwarding information inquiry to the home location register may be made before the optimal routing decision in the GMSC, if this has been indicated in the original routing information response by the home location register; the details are not described herein.

If the gateway MSC cannot resume call handling, call forwarding is handled by the serving VMSC-B according to the current procedures.

5. DISCUSSION

This chapter has given an overview of the CAMEL and optimal routing features introduced to the GSM system in phase 2+ standardization. The features are generally expected to be available from equipment vendors in late 1998 or 1999, but the estimate is of course subject to uncertainties concerning the final acceptance of the specifications in ETSI. Another matter is then when the features will be in use in networks. Both features become fully operational, i.e. benefit roaming subscribers, only when they are available in a large enough base of networks.

Both CAMEL and optimal routing will have an impact on the number of HLR transactions and signalling traffic. The main reason for the increase in the HLR load is likely to be the terminating CAMEL. A call to a subscriber with terminating CAMEL services requires two HLR interrogations whereas only one interrogation suffices without CAMEL, assuming that CAMEL does not modify the call destination. If a certain percentage of the subscribers in a GSM network subscribes to terminating CAMEL services, the number of HLR transactions due to terminating calls could increase by the same amount. Also optimal routing affects the HLR load. First, an extra interrogation occurs when optimal routing of a mobile-to-mobile call is attempted, but the home PLMN route must eventually be used. Second, the criteria of optimal routing are checked by HLR, which increases its processing load somewhat.

Signalling load in the SS7 network will grow because of the increased number of interrogations, and especially, because of the gsmSSF-gsmSCF dialogue in the CAMEL feature. Note that these signalling connections are international for subscribers roaming in foreign countries. Also the amount of data transferred with GSM MAP in the SS7 network increases due to the CAMEL information included in the subscriber data, but the effect of this will probably be minor. The increased signalling load due to optimal routing interrogations is naturally balanced by the improved efficiency in using landline trunks.

ACKNOWLEDGEMENTS

The author would like to thank an anonymous reviewer, the editors, and his colleagues for comments that have greatly improved the presentation.

ABBREVIATIONS

ACM	Address Complete Message
CAP	CAMEL Application Part
DP	Detection Point
EIR	Equipment Identity Register
GMSC	Gateway MSC
gsmSCF	GSM Service Control Function
gsmSSF	GSM Service Switching Function
HLR	Home Location Register
IAM	Initial Address Message
IMSI	International Mobile Subscriber Identity
IN	Intelligent Network
INAP	Intelligent Network Application Protocol
ISUP	ISDN User Part
MAP	Mobile Application Part
MSC	Mobile Switching Center
MSISDN	Mobile Subscriber ISDN number
O-CSI	Originating CAMEL Subscription Information
PLMN	Public Land Mobile Network
PRN	Provide Roaming Number
PSI	Provide Subscriber Information
RCH	Resume Call Handling
REL	Release
SIFIC	Send Info for Incoming Call
SIFOC	Send Info for Outgoing Call
SRI	Send Routing Information
SS7	Signalling System No. 7
TCAP	Transaction Capabilities Application Part
T-CSI	Terminating CAMEL Subscription Information
VLR	Visitor Location Register
VMSC	Visited (Serving) MSC

REFERENCES

Ali-Vehmas, T. (1997). "Smart phones – Designed to connect people," in the 1997 GSM World Congress, Cannes, France, February 19-21, 1997.

ETS 300 374-1. Intelligent Network (IN); Intelligent Network Capability Set 1 (CS1) Core Intelligent Network Application Protocol (INAP) Part 1: Protocol specification. ETSI, 1994.

GSM 02.78. Customized Applications for Mobile network Enhanced Logic (CAMEL); Service definition (Stage 1). ETSI, GSM Technical Specification 02.78, Version 5.1.1, March 1997.

GSM 02.79. Support of Optimal Routeing (SOR); Service definition (Stage 1). ETSI, GSM Technical Specification 02.79, Version 5.0.0, July 1996.

GSM 03.78. Customized Applications for Mobile network Enhanced Logic (CAMEL) - Stage 2. ETSI, GSM Technical Specification 03.78, Version 5.0.0, April 1997.

GSM 03.79. Support of Optimal Routeing (SOR); Technical Realisation. ETSI, GSM Technical Specification 03.79, Version 5.0.0, April 1997.

GSM 09.02. Mobile Application Part (MAP) specification. ETSI, GSM Technical Specification 09.02 (ETS 300 599), November 1996.

GSM 09.78. CAMEL Application Part (CAP) specification. ETSI, GSM Technical Specification 09.78, Version 5.0.0, April 1997.

Q.1201. Principles of intelligent network architecture. ITU-T Recommendation, October 1992.

UMTS 22.01. Universal Mobile Telecommunications System (UMTS); Service aspects; Service principles, Version 3.1.0. ETSI, June 1997.

Verkama, M., Söderbacka, L., and Laatu, J. (1996). "Mobility management in the third generation mobile network," in Proceedings of IEEE Global Telecommunications Conference, Vol. III, pp. 2058-2062, London, November 18-22, 1996.

PART 2

RADIO ASPECTS

Chapter 1

Antenna Arrays and Space Division Multiple Access

Preben E. Mogensen, Poul Leth Espensen, Klaus Ingemann Pedersen, Per Zetterberg[1]
Center for Personkommunikation, Aalborg University, Denmark

Key words: adaptive antenna arrays, space-division multiple-access, diversity, combining.

1. INTRODUCTION

Adaptive base station antennas - also called *smart antennas* have recently become a *hot* research topic. In principle the adaptive antenna concept consists of an antenna array and a flexible beamforming network.

In terms of commercial exploitation there are mainly two driving factors for introducing adaptive base station antennas, (i) range extension, (ii) capacity enhancement.

The issue of range extension is briefly discussed in Section 2, but the main emphasis is put on capacity enhancement of adaptive base station antennas. As sketched in Figure 1, an adaptive antenna array typically creates a narrow beam in the direction of the desired mobile station and thereby reduces the co-channel interference level in other azimuth directions. This is called spatial filtering. There are two methods to exploit a capacity gain from spatial filtering: In the concept of Same Cell Reuse (*SCR*), the physical radio channels (timeslot, frequency channel) = (t,f) are

[1] Visitor from KTH, Sweden. He is currently working at Radio Design in Stockholm.

being reused several times within a cell. The other concept is Reduced Cluster Size (*RCS*). The *RCS* concept in a combination with random Frequency Hopping (FH) is found most suitable for GSM.

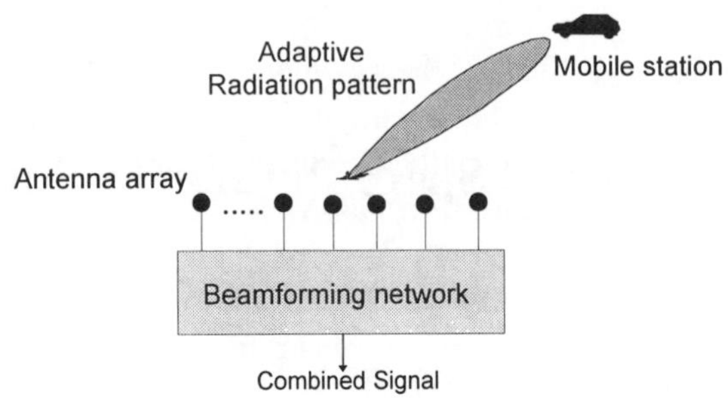

Figure 1. Adaptive base station antenna array concept.

The spatial filtering gain of an adaptive antenna array is obviously related to the size and the number of elements in the array, but it is also strongly related to the azimuth spread of the multipath propagation channel. Basic information on the azimuth channel characteristics has been included, because this is found essential for understanding the *real* challenge of introducing adaptive antennas in mobile communications. Implementation technology of the beamforming network and algorithms design are also important issues. The access technology and other system parameters have a great impact on how complex it is to exploit the potential capacity enhancement of adaptive antennas. The aim is to give the reader a broad overview on the performance and complexity of the use of adaptive antennas in GSM[2].

The authors have participated in the European TSUNAMI II[3] research project on adaptive antennas for GSM/UMTS. The presented results and views are mainly based on the authors' own results from this research project. Other ongoing research projects on adaptive antennas for GSM are [For94, And97, Tan97, Rat96] among others.

[2] There will only be distinction between GSM-900, DCS1800, and PCS1900 when considering the carrier frequency. Otherwise GSM will be used to cover all three systems.

[3] TSUNAMI II is a European research project on adaptive base station antenna arrays funded by the European Commission under the ACTS programme.

2. RANGE EXTENSION

In the evolution of mobile communication systems, there has been a continuous drive of movement to the higher frequency bands, in order to obtain more frequency spectrum. In Europe, for example, only 2x25MHz is available for GSM-900 services, whereas a band of 2x75MHz is reserved for DCS-1800. Measurement results presented in [Mog90, Mog91] have shown that the path loss in urban areas is approx. 10 dB[4] higher at 1800MHz compared to 900MHz. This additional path loss significantly increases the cost of rolling out a DCS-1800 network compared to a GSM-900 network. Base station antennas with a high horizontal (and also vertical) directivity can be used to compensate for the increased propagation path loss. There are mainly two network solutions for installing base station antennas with a high horizontal directivity, i.e., small beamwidth:

- Use of a more narrow sectorization than 120°. However, such a solution has limited potential and some drawbacks in terms of an increased handover rate and a reduced trunking efficiency.
- The better alternative solution is to use an *adaptive* narrow beam antenna system, which can find and track the azimuth direction to the mobile stations within a sector. Such a solution can be implemented by switching between several narrow beam antennas having different azimuth directions or by using an antenna array with a flexible (adaptive) beamforming network. Only the adaptive antenna array technology will be addressed further.

The azimuth array factor (AF) gain of an antenna array with M horizontal elements is defined as: $10\log(M)$ in dB. Assuming a simple power loss model, with $r^{-\gamma}$ being the power loss, and r being the distance, then the range extension over a single antenna element is given by $M^{1/\gamma}$. For example, for a path loss slope of $\gamma = 3.5$ and M=4, the coverage range is increased by a factor 1.5 over a single element. In Table 1 the azimuth AF gain and the horizontal size of an antenna array are given for various M values:

4 The path loss difference is expected slightly lower for urban rural areas

M	AF gain (dB)	Range extension	Beam width deg.	Size (m) 900 MHz	Size (m) 1800 MHz
4	6	1.5	26	0.7	0.4
8	9	1.8	13	1.4	0.7
16	12	2.2	6.4	2.7	1.4

Table 1. Potential range extension and horizontal size of an antenna array. The assumed element spacing is $\lambda/2$, λ being the wavelength of the carrier frequency .

It should be noticed that the physical size of the antenna array (for environmental reasons) sets an upper limit for the number of elements, thus allowing more elements at an 1800MHz carrier frequency than for 900MHz. The values listed in Table 1 are under ideal circumstances, meaning omni-directional elements, no mutual coupling between elements, no scanning loss, and that all multipath signal components are captured by the beam. The influence of azimuth multipath spread is discussed later in sections 5.

3. CAPACITY ENHANCEMENT METHODS

The potential capacity enhancement of using adaptive base station antennas is achieved by a spatial filtering gain, i.e., reduced co- (and adjacent) channel interference in the network.

Two different network strategies can be used to convert the spatial filtering gain of adaptive antennas into a capacity gain: (i) Reduced Cluster Size (*RCS*), and (ii) Same Cell Reuse (*SCR*). For *RCS* mode the spatial filtering gain is used to reduce inter-cell co-channel interference. Hence the cluster size[5] can be reduced without sacrificing the link quality. Again, assuming a simple path loss model, $r^{-\gamma}$, and a spatial filtering gain proportional to the number of elements *M*, then the relative reduction in cluster size, $\Delta K = {}^{K_{RCS}}\!/_{K_{Sector}}$ can be expressed as: $\Delta K = M^{-2/\gamma}$, where K_{RCS} is the cluster size using adaptive antennas, and K_{Sector} is the minimum cluster size with a conventional sector base station antenna. It here assumed that the inter-cell co-channel interference is uniformly distributed in azimuth and the antenna beamwidth is inverse proportional to the number of elements (see Table 1). For M=8 and γ=3.5 this yields ΔK=0.3, which is equivalent to an improved spectrum efficiency of a factor 3.3. The

5 The cluster size determines the number of cells in cell-group that do contain co-channel interference. For hexagonal cell shape the cluster size K is given by $D^2/(3R^2)$, where R is the cell radius and D is the co-channel cell separation [Meh94].

concept of using an adaptive narrow beam antenna to reduce inter-cell co-channel interference has been described in the literature as spatial filtering for interference reduction (SFIR) [Tan95].

SCR mode is in the literature also often called Space division multiple access (SDMA). In *SCR* mode several mobile stations simultaneously use the same physical channel within a cell. If the position of the mobile stations, which share a physical channel, are located differently in azimuth, then the signals can be sufficiently separated to perform successful demodulation by means of spatial filtering. A third dimension for user separation is introduced; We now have frequency, time, and azimuth. The capacity increase of SDMA mode is equal to the expectation value for the number of users that can share the same physical channel (*t,f*). The capacity gain of *SCR* mode is more complicated to analyse than for *RCS* mode, because of the near-far problem, when having non-ideal uplink RF-power control. This issue is discussed in more detail in Section 10. The principles of both the *SCR* and *RCS* modes are depicted in Figure 2.

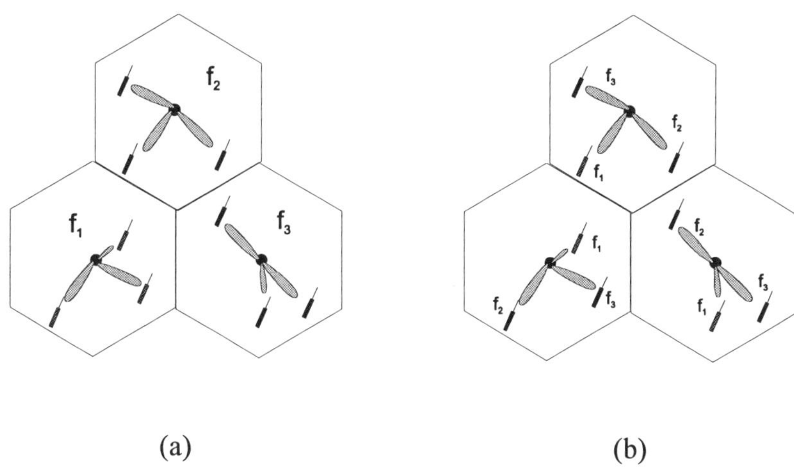

Figure 2. (a) Principle of SCR mode where three co-channel SDMA users are sharing the same physical channel (t,f) and the cluster size is 3. (b) principle of RCS, where the cluster size is 1, and therefore 3 times as many physical channels are available per cell. For the examples (a) and (b) the network capacity of the SCR and RCS modes is shown to be identical. This may not be true in reality.

4. DIRECTIONAL CHANNEL MODELS

The effective gain (whether capacity or range extension) of adaptive antennas is closely related to the azimuth spread of the multipath propagation channel. Performance evaluation of adaptive antenna arrays by means of theoretical investigations or Monte-Carlo simulations only yields realistic results provided that the underlying multipath propagation model accurately reflect the statistics of the *real* radio channel. In the following we give a simple azimuth characterisation of the multipath radio channel, which is the basis for the discussion on effective radiation pattern in section 5.

When introducing adaptive antennas in mobile communications it becomes insufficient to only characterise multipath propagation in the temporal domain - the spatial domain now becomes equally important. Characterisation of radio channels in the temporal domain has been investigated intensively in numerous studies, where some of the first references are [Bel63, Tur72], whereas only recent research have focused on the spatial characterisation [Egg95].

We start by defining the two-dimensional complex channel impulse response as $h(\tau,\theta)$, where τ is the delay, and θ is the azimuth. Based on this definition, the Power Azimuth Spectrum (PAS) is defined as

$$P(\theta) = \left\langle \int \left| h(\tau,\theta) \right|^2 d\tau \right\rangle \quad \text{for } \theta \in]-\pi,\pi] \tag{1}$$

where $\langle\cdot\rangle$ denotes local average over short term fading. Eggers introduced the quantity *Azimuth Spread* (AS) [Egg95] being defined as the standard deviation of the PAS. The AS of the power azimuth spectrum is thus equivalent to the RMS delay spread of the power delay profile. The PAS seen by a macro-cellular base station antenna has been found to be accurately modelled by a Laplacian function [Ped97a]:

$$p(\theta) = \frac{k}{\sqrt{2}\sigma_\theta} \exp\left(-\sqrt{2} \frac{|\theta - \mu_\theta|}{\sigma_\theta} \right) \tag{2}$$

where σ_θ is the AS, μ_θ is the mean azimuth direction, and k is a scaling factor equal to one for unit received power. This simple model applies well for both urban and rural areas. An example of a measured PAS in a rural area is shown in Figure 3. For this particular situation the AS is approx. 2°, meaning that the impinging power to the base station antenna array is very concentrated around the mean direction μ_θ, which is the direction to the mobile station.

The AS is a stochastic quantity, which is strongly correlated with the shadow fading component [Mog97]. Measurement results from the city of Aalborg showed that the AS typically increases by 1° as the local signal strength fades 10 dB [Mog97]. Figure 4 shows the cumulative distribution of the AS from measurements in a dense urban area for three different base station antenna heights [Ped97b]. The lowest antenna height in Figure 4 is equivalent to rooftop level of adjacent buildings, whereas the two other antenna heights correspond to approx. 6 meters and 12 meters above rooftop level, respectively. The scenario for the two highest antenna positions can be considered as typical urban macro cellular situations, while the low antenna position is in the transition zone between a macro and a micro cell installation. It can be observed that the AS significantly increases as the antenna height is reduced. At the 50 percentile the AS raises from 5° to 10°, when moving from the highest to the lowest antenna position. The mean path loss was at the same time increased by 7 dB.

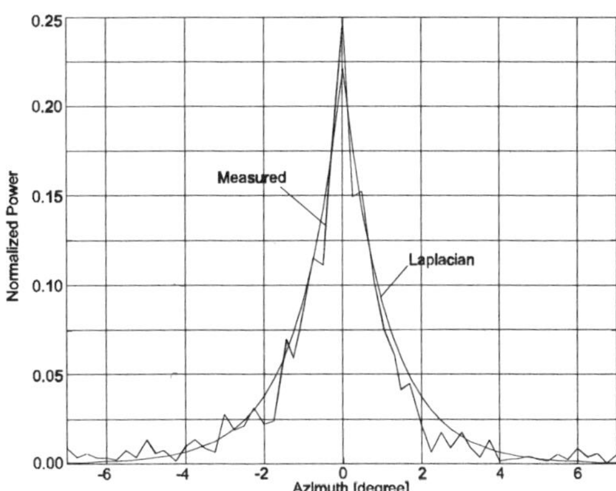

Figure 3. Power azimuth spectrum (PAS) for a rural area; Measured and modelled.

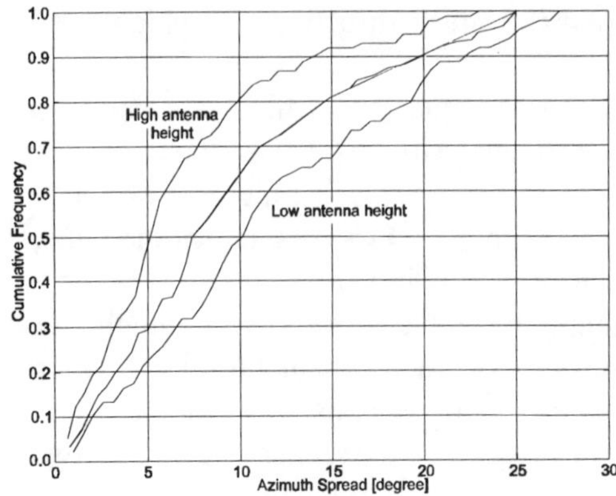

Figure 4. Cumulative frequency distribution of azimuth spread (AS) in an urban environment (Århus) for three different antenna heights: 32m, 26m, and 20m .

Although the PAS is accurately modelled by a Laplacian function for most measured routes, there exist special cases, where the shape is very different. Figure 5 illustrates such two situations. The PAS in Figure 5a exhibits a narrow peak at zero degrees plus an almost uniform power distribution in the range 0°-12°. The PAS in Figure 5b looks Gaussian with mean at -6°. These special cases were very rarely observed but may likely happen more frequently for less homogeneous city areas (i.e. urban areas with rivers, hills or a large number of high rise buildings).

Another interesting observation from the measurements performed within the TSUNAMI II project was that there is no simple relation between AS and distance as suggested in the literature, for example [Lee73]. For rural areas, measurement results from both Denmark and UK have shown both decreasing and increasing AS versus distance. The AS is more effected by the local *features* of the environment than the distance between the base- and the mobile station. Results for urban areas indicate that the AS is almost invariant with distance (at least in the measured range of 0.5-3km).

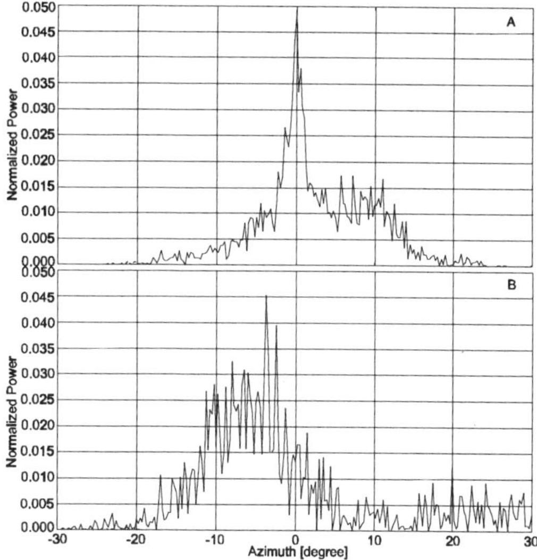

Figure 5. Two examples of atypical power azimuth spectrum.

5. EFFECTIVE RADIATION PATTERN

The effective radiation pattern of an adaptive antenna array is distorted by the power azimuth spectrum (PAS), when the antenna beamwidth θ_{3dB} becomes narrow relative to the AS σ_θ. A large AS severely reduces the spatial filtering gain of transmitting (or receiving) with a narrow antenna beam. Opposite, a large AS is an advantage for horizontal space diversity reception (or transmission), since it improves the de-correlation of signals received on each element in the antenna array. Figure 6 shows the signal envelope correlation as a function of the element spacing assuming a Laplacian PAS.

In the following we introduce the *effective radiation pattern* of a narrow beam antenna taking the PAS into account. The M elements in the linear antenna array are assumed combined by means of conventional beamforming [Veen88], see Figure 7.

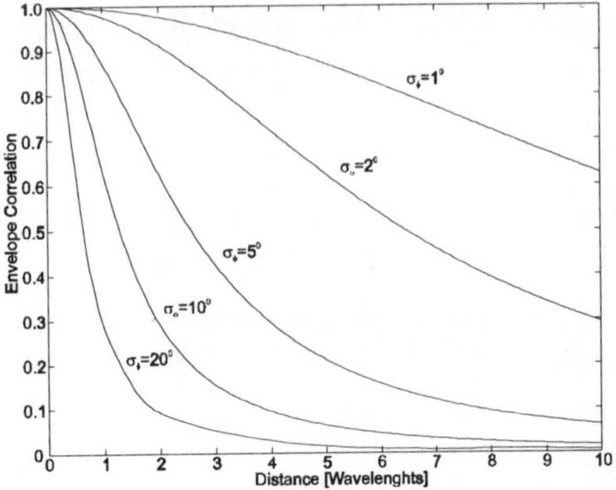

Figure 6. Envelope correlation as a function of the horizontal antenna element spacing for various azimuth spread values σ_θ. A Laplacian power azimuth spectrum is assumed.

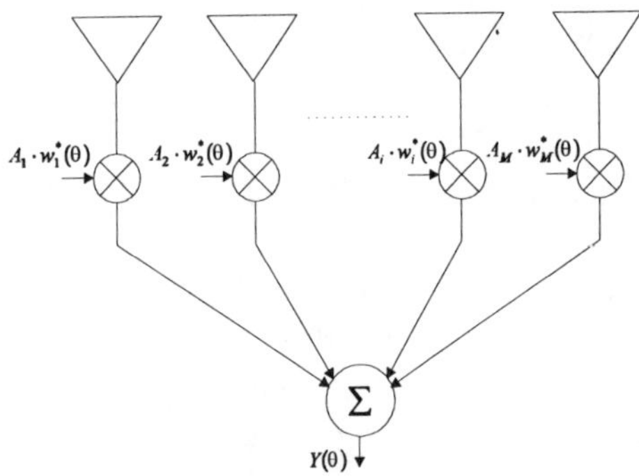

Figure 7. Linear beamforming network.

The complex weights $w_i(\theta)$ in Figure 7 represent the phase shift from element number one to element number i.. In the literature, the vector $\boldsymbol{w}(\theta) = [w_1(\theta) \;\; w_2(\theta) \;\; \ldots\ldots\ldots \;\; w_M(\theta)]^T$, is often referred to as the array steering vector. The complex response of the antenna array, with a steering direction θ_0 and a point source in the direction θ, can thus be written as

$$Y(\theta,\theta_0) = \sum_{i=1}^{M} A_i \cdot w_i(\theta) \cdot w_i^*(\theta_0) \tag{4}$$

where A_i is an amplitude weight, which may be selected according to a window function. For a PAS given by a distribution function $p(\theta)$ - Eq. 2

for a Laplacian - impinging with the mean azimuth direction θ_1, the received average power, when steering the beam towards the direction θ_0, is given by

$$P(\theta_1, \theta_0) = \int \left| Y(\theta, \theta_0) \right|^2 \cdot p(\theta - \theta_1) d\theta \tag{5}$$

Figure 8 shows the received power versus the steering direction θ_0, for different values of the σ_θ. The mean azimuth direction θ_1 of the Laplacian PAS is set to zero degrees.

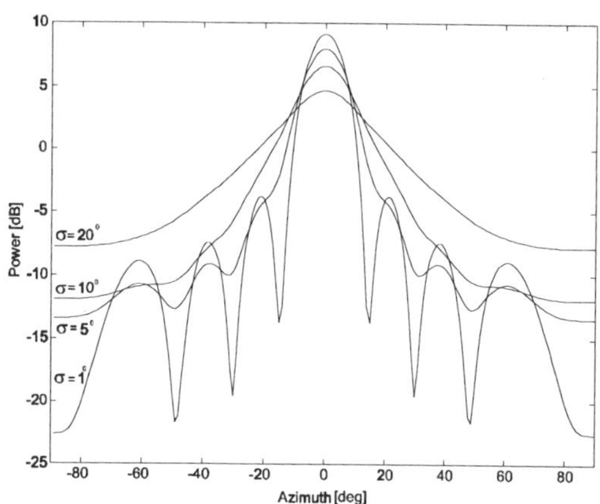

Figure 8. Received power of an 8 element uniform linear array with an element spacing of λ/2 as a function of the steering direction θ_0 for a Laplacian PAS having a mean direction of $\theta_1=0^0$. The received power is shown for different azimuth spread values, σ_θ.

By applying a non-uniform window function it is possible to suppress the sidelobe level at the expense of a broader mainlobe and a reduced gain. For example, using a Kaiser window function the width of the mainlobe and the sidelobe suppression can be controlled in a flexible manor. The weights of the Kaiser window are given by:

$$A_i = \frac{I_0\left(\alpha \cdot \sqrt{1 - \left(\frac{i}{N}\right)^2}\right)}{I_0(\alpha)} \tag{6}$$

where the value α controls the shape of the Kaiser window. In Figure 9 the received power P for different angles θ_0 and for $\sigma_\theta = 5^\circ$ is shown for different values of α.

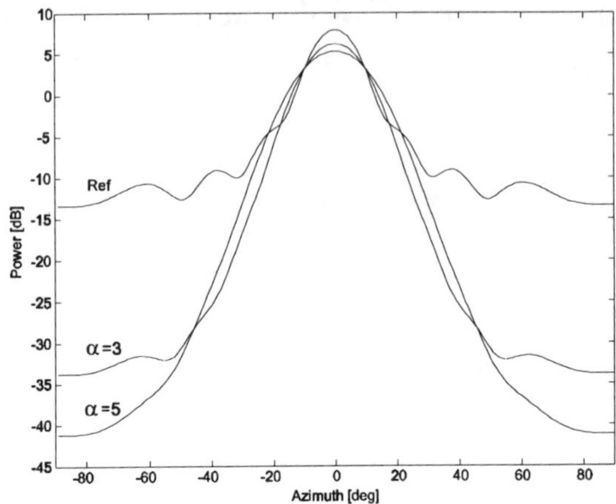

Figure 9. Received power of an 8 element linear array with an element spacing of $\lambda/2$ as a function of the steering direction θ_0 for a Laplacian PAS having a mean direction of $\theta_1=0°$ and $\sigma_\theta=5°$. The parameter α in the Kaiser window function (see Eg. 3.1.6) is used to control the side lobe level.

When two or more mobile stations in SDMA mode share the same physical channel (t,f), it is important to know the spatial filtering gain as a function of the azimuth separation. If the bcamforming network is set to steer in the direction θ_0, which is the direction of the desired mobile station, and interference is coming from a mobile in the direction θ_1, then the carrier to interference ratio can be computed as: C/I = $P(\theta_0, \theta_0)/P(\theta_1, \theta_0)$. For simplicity it is here assumed that the received signal strength for the two mobile stations is identical, otherwise the calculated C/I value should be adjusted by the difference in signal strength.

Figure 10a shows the function $P(\theta_1, \theta_0)$ for a desired mobile station located in the direction $\theta_0=0°$, and the interfering user at $\theta_1=30°$. The C/I for this situation can be found to be approx. 17dB. In Figure 10b the C/I is shown as a function the azimuth direction θ_1 of the interfering user. The curves are shown for three different locations of the desired user (θ_0). It can be observed that for a linear antenna array, the required azimuth separation, in order to obtain a certain C/I, increases when the desired mobile station moves towards endfire.

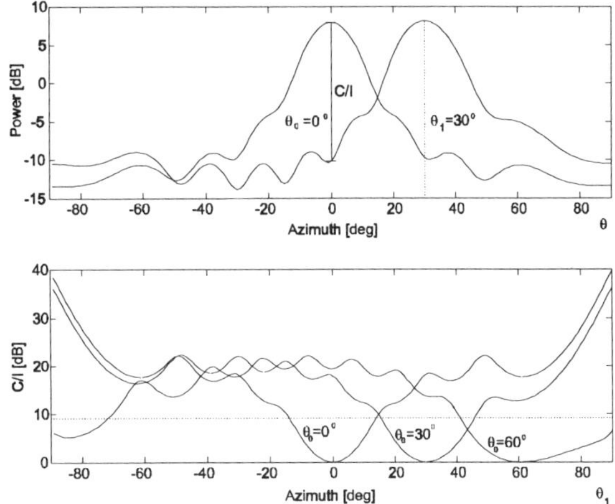

Figure 10 (a) C/I for two SDMA users separated by 30°. (b) C/I as a function of the position θ_1 of the interfering user for 3 positions θ_0 of the desired user.

It can be observed from Figure 8 that for large AS the AF gain decreases. When the AS become large relative to the antenna beamwidth not all the impinging waves will be captured, and the effective AF gain saturates [Nor94]. In Figure 11 the relative loss in captured signal power is shown as function of the 3 dB beamwidth (i.e., *M)* for different AS values. The curves are computed analytically, assuming a Laplacian PAS. The *ideal* AF gain is also shown in the Figure (dashed line). For example, a 16-element antenna array has an ideal AF gain of 12dB and a 3dB beamwidth of approximately 6.4°. For an AS value of 10° the captured power of an 16 element array is reduced by 4.5dB relative to a point source. Hence, the effective AF gain is only (12-4.5)dB = 7.5dB. Similarly the effective AF gain for *M*=8 can be found to 6.75dB for an AS of 10°. This means that the effective gain increasing from *M*=8 to *M*=16 is less than 1 dB.

In urban environments with large AS, only a minor improvement is achieved by using antenna arrays with more than 6-8 horizontal elements. In rural environments with low AS (2°-4°), almost the full theoretical AF gain is achieved up to *M*=12. Measurement results from a rural area presented in [Tan97], using an 8 element linear array, have verified that the theoretical AF gain of 9dB can be achieved. It may however, be an advantage to use more antenna elements than discussed above in order to apply a non-uniform window function for improved sidelobe suppression, or if using a transmit or receive algorithm different from beamsteering.

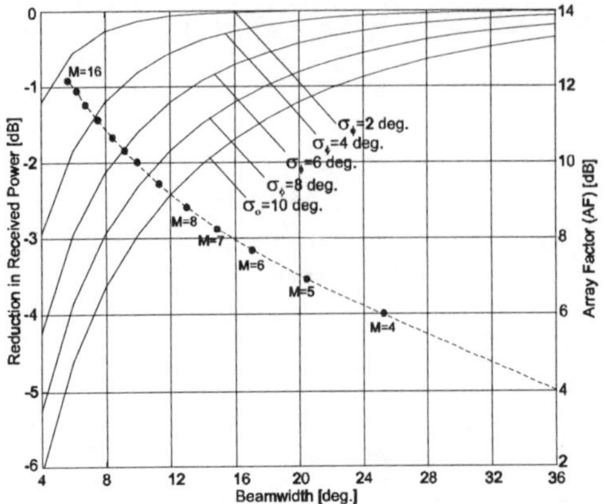

Figure 11. Relative signal power loss versus antenna beamwidth for different azimuth spread.

6. ANTENNA ARRAY AND CELL TOPOLOGY

A number of different antenna array topologies have been analysed in [Mog97a]. There are mainly two categories for a network layout in GSM to be considered: A sectorized cell configuration, and an omni-cell configuration. For the omni-cell configuration, it is recommended to use a circular antenna array (or an approximation to a circular array) in order to avoid discontinuities in azimuth. Omni-cells are probably the preferred network configuration for *SCR* mode (this is discussed further in section 10. One potential drawback of a circular antenna array is a less flexible installation compared to a sectorized antenna configuration.

Today, a three sectorized base station configuration is most commonly used for a GSM macro-cellular base stations. Such a network layout is probably most suited for the *RCS* mode and may easily use a linear antenna array to cover each sector [Mog97a].

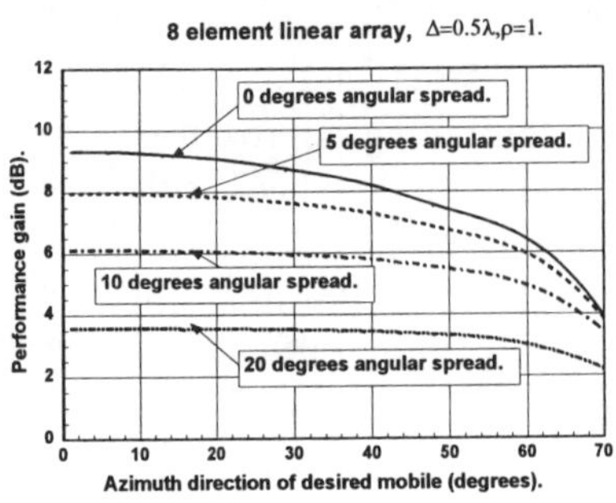

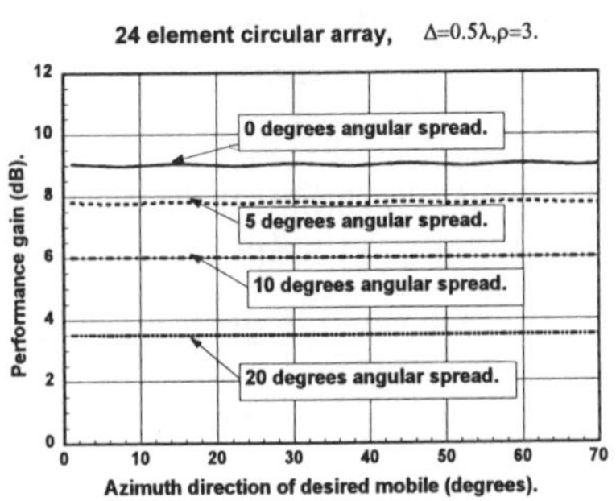

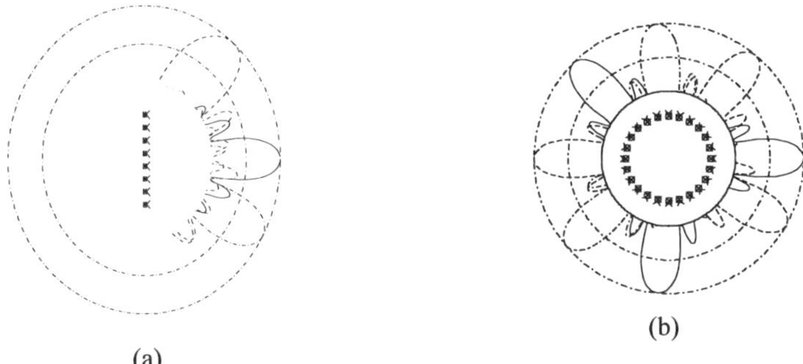

(a)

(b)

Figure 12. (a) Performance improvement of an 8 element linear array to cover a sector of 120°. (b) performance of a 24 element circular array to provide omni-directional coverage. Δ is the element spacing in λ, and ρ determines the element directivity: $p(\theta)=\cos^{\rho}(\theta)$ for $-90<\theta<90$, and zero otherwise.

In Figure 12(a) the C/I gain of an 8 element linear array (three sectored BS), and in (b) a 24 element circular array (omni-sectored BS) is shown for the beamsteering approach. The performance of the linear array has an azimuth dependency: For an AS of 5° the C/I performance has been computed to be 8dB at broadside direction, which falls off to 6 dB at 60° off broadside direction. By using a network layout, where the base station sites are positioned in the corners of a hexagonal grid, the density of mobile stations at ±60° is only half of the density at broadside. The reduced gain at ±60° is therefore not problematic. For the circular array the C/I gain is close to 8dB for an AS of 5°, an azimuth invariant. Co-channel interference is assumed uniformly distributed in azimuth for the results in Figure 12.

7. BEAMFORMING TECHNOLOGY

The beamforming network of an adaptive antenna can be implemented in either analogue RF circuitry or in fast digital signal processing hardware (at baseband). Both implementations have pros and cons. The main advantage of the baseband digital signal combining approach is a higher flexibility in radiation pattern synthesis. The main drawback is the demand for *on-line* calibration to compensate for dissimilar transfer functions of each receiver/transmitter chain [Mog96]. RF beamforming for a linear antenna array may use a Butler matrix implementation [She61]. The Butler matrix does not include any active circuitry and does therefore not demand on-line calibration. A second advantage of using a Butler matrix is reduced requirements for high dynamic range receivers, because the spatial filtering is performed before amplification. Drawbacks of using a Butler matrix implementation are, however, the insertion loss of typically 0.5-1 dB [Ana], and a scanning loss of approx. 4 dB in between two adjacent beams. In [Joh97] a compact 6 element antenna array configuration with slanted ±45° polarised elements and a separate 6x6 Butler matrix for the two polarisations have been presented. The beam directions for the two polarisations are interleaved in order to reduce the cross over depth between adjacent beams to only about 1dB. Uplink reception with a Butler matrix beamformer does not prevent implementation of optimum combining or any other diversity scheme in base band signal processing, as it is just a linear device.

Transmission with adaptive antennas is more complex than reception due to the much higher RF power level. The power amplification of multi-carrier or SDMA signals can be performed either before or after the signal combining. From an integration point of view, it would be preferable to combine all the GSM multi-carrier or SDMA signals before power amplification, but this requires extremely linear power amplifiers (PA's). Within the near future it is probably unrealistic to combine and amplify multi-carrier signals, which in principle can span over a frequency range of 75 MHz in DCS-1800. For SDMA signals, which have the same carrier frequency, it is easier to meet the intermodulation product requirements of GSM. For multi-carrier SDMA configurations, a hybrid solution may therefore be used. If performing power amplification before multi-carrier signal combining, there is a need for M x K PA's, where M is the number of antenna elements (or the number of beam directions when using a Butler matrix) and K is the number of carrier frequencies. For a base station with low transmit power - using a Butler matrix for beamforming, it may be possible to perform the beam switching after power amplification. This reduces the number of PA's to only K (the same number as without adaptive antennas). It is important to notice that the PA output power can be reduced nearly by a factor M for the same coverage as without adaptive antennas (insertion and scanning loss as well as azimuth spread are to be considered).

8. TRANSMIT AND COMBINING ALGORITHMS

GSM uses Frequency Division Duplexing (FDD), and therefore there is a major difference between uplink reception and downlink transmission with adaptive antennas. For uplink reception, the instantaneous state of the multipath radio channel can be estimated from the training sequence, and the instantaneous co-channel interference and noise can be estimated. Linear types of diversity combining schemes, including optimum combing, can therefore easily be applied [Win94]. Maximum likelihood sequence estimation (MLSE) can also be used for joint array processing and demodulation [Bot95].

For downlink transmission the situation is less fortunate. The instantaneous state of the multipath fading channel is unknown at the base station beamformer. The duplex distance is 45MHz for GSM-900 and 95MHz for DCS1800, which is far beyond the coherence bandwidth, and therefore the instantaneous impulse response at the downlink frequency can not be extracted from the uplink signal. For the *RCS* mode in a combination with random FH, it is also impossible to estimate the downlink co-channel interference level at the mobile station. Most research activities on algorithm design for adaptive antennas have been addressing the uplink combining case, despite the conditions for downlink transmission is less fortunate and will therefore likely be the capacity limiting link direction. A simple and straight forward way of implementing downlink beamforming is to transmit in the azimuth direction of the desired mobile station (or actually in the direction where the dominant signal path averaged over short term fading is received). However, this is a sub-optimal solution since the direction of co-channel users are not taken into account. A thorough analysis of downlink performance with downlink nulling for both *RCS* and *SCR* can be found in [Zet97].

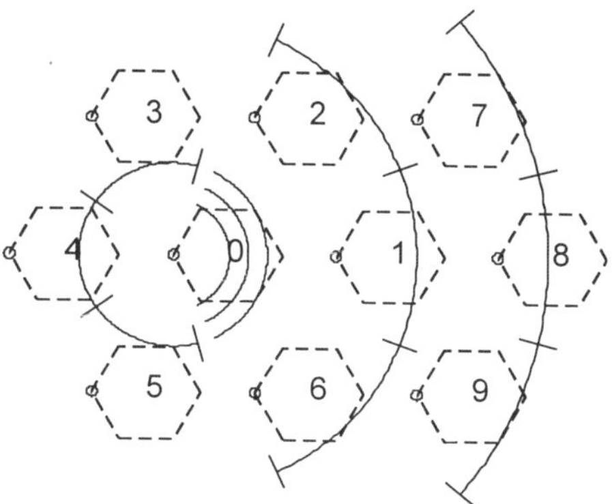

Figure 13. Cellular geometry of model used for simulations in a GSM network. The simulations include test of Direction of Arrival (DoA) algorithms and uplink diversity combining algorithms. The desired user follows arcs in cell #0 while the interfering users move on the other shown arcs.

In the following sections we look into design and performance of a beamsteering algorithm for downlink transmission and an optimum combining algorithm for uplink reception.

A simulation model of a GSM network using a frequency reuse 1/3 has been developed in order to test algorithm performance [Mog97a]. The network layout in the simulator is shown in Figure 13. The simulator can import both measured and simulated radio channel data for the desired user in cell #0. The co-channel interfering users in cell #1-9 do always use simulated radio channel data. The power delay profile of the simulated radio channel is made according to the GSM TU profile (RMS delay spread of 1μs), and the power azimuth spectrum is Gaussian[6] with an AS of 5°. A pathloss model of $r^{-3.5}$ and 8 dB log-normal fading is used in the simulations. GSM features, such as DTX and FH are included, and the network can be configured to both synchronous and asynchronous mode.

8.1 Transmission

For downlink transmission one would ideally like to know the multidimensional impulse response matrix $\mathbf{H}_b^{TX}$ for the burst number b. This matrix contains the impulse response for each antenna element in the base station antenna array and has dimension M x N where N is the length of the temporal impulse response.

$$\mathbf{H}_b^{TX} = \left[\mathbf{h}_{1,b}^{TX}\ \mathbf{h}_{2,b}^{TX} \ldots \mathbf{h}_{s,b}^{TX} \ldots \mathbf{h}_{N,b}^{TX}\right] \tag{7}$$

The vector $\mathbf{h}_{s,b}^{TX}$ is the spatial impulse response vector for the access delay index s. The dimension of $\mathbf{h}_{s,b}^{TX}$ is M x 1. Unfortunately, it is not possible to estimate this impulse response. However, under the assumption of Gaussian Wide Sense Stationary Uncorrelated Scattering (GWSSUS) channel model, the expectation value for $\mathbf{h}_{s,b}^{TX}$ can be found from uplink reception.

Since the received signal quality, at the outage threshold, is almost a linear function of the received signal strength in dB, the following performance criterion function

6 A Gaussian PAS was initial expected, before measurement results showed a better fit to a Laplacian function.

$$\mathbf{w} = \arg_{\mathbf{w}} \max\left\{ \mathrm{E}\left\{ 10\log\left(\sum_{s} \left| \mathbf{w}^{*}\mathbf{h}_{s}^{TX,0} \right|^{2} \right) \right\} - 10\log\left(\mathbf{w}^{*}\mathbf{w} \right) \right\} \tag{8}$$

is used for computing the transmit weights **w** (index 0 corresponds to cell #0 in the simulation model, see Figure 13). The expectation, $E\{\ \}$ is performed over the fast fading.

In [Mog97a] it is argued that for typical radio channel conditions less than one dB is lost by restricting the transmit weighting vector to the form

$$\mathbf{w} = const \times \mathbf{a}^{TX}\left(\theta_{0}, f^{TX}\right) \tag{9}$$

where f^{TX} is the downlink frequency and θ_0 is the steering direction, and

$$\mathbf{a}(\theta, f) = \left[1, \exp(-j2\pi f\Delta\sin(\theta)/c), \ldots, \exp(-j(M-1)\Delta\sin(\theta)/c)\right]^{T} \tag{10}$$

is the array steering vector, Δ is the antenna element spacing and c is the speed of light. This constraint on **w** reduces the degrees of freedom to only one. By inserting Eq. (9) into Eq. (8) and simply replace TX by RX[7]. yields:

$$\theta_{0} = \arg_{\theta} \max\left\{ E\left\{ 10\log\left(\sum_{s} \left| \mathbf{a}^{RX}\left(\theta, f^{RX}\right)^{*} \mathbf{h}_{s,b}^{RX,0} \right|^{2} \right) \right\} \right\} \tag{11}$$

which is the selected downlink transmit direction. This algorithm can also interpreted as a *low resolution* Direction of Arrival (DoA) estimator. A practical implementation of this DoA algorithm for GSM is described in the following.

DOA ESTIMATION ALGORITHM FOR GSM

The GSM traffic burst includes a midamble of 26 known symbols $I^{RX,0}$ (the training sequence). The impulse response estimates $\hat{\mathbf{h}}_{s,b}^{RX,0}$ are obtained by a cross-correlation between the received signal and a MSK mapped version of the known training sequence (16 or 26 bits may be used).

[7] It can be shown that replacing TX by RX can be done if the element pattern for up- and down-link are similar.

$$\hat{\mathbf{h}}_{s,b}^{RX} = \sum_{q=s}^{s+25} \frac{1}{26} \mathbf{x}_{q,b} I_{q-s,b}^{RX,0*} \tag{12}$$

where $\mathbf{x}_{q,b}$ is the spatial signal vector for symbol q in burst b and * denotes complex conjugation. For each antenna element, the estimated impulse response is N=11 symbols long (±5 symbols timelag), and the dimension of $\hat{\mathbf{H}}_{s,b}^{RX,0}$ becomes M x 11. For the case of an eight element uniform linear array covering a sector of 120°, it was selected in [Mog97a] to have a cross-over depth of 0.5 dB between adjacent beams, in order to reduce the computational load. This gives the 22 fixed beam directions:

$$\Theta = [-72.7,-59.4,-49.8,-41.9,-35.0,-28.5,-22.6,-16.6,-11.0,$$
$$-5.5,0.0,5.5,11.0,16.6,22.6,28.5,35.0,41.9,49.9,59.4,72.7,88.2]$$

For equalisation in GSM it is typically assumed that the Power Delay Profile (PDP) is shorter than 5 symbols. In order to reduce the impact of co-channel interference on the DoA estimation, it is here further assumed that most of the energy in the PDP for the desired transmit direction θ_0 is kept within a period of only 3 symbols $s \in \{k_0, k_0+1, k_0+2\}$ where k_0=1,..9. The time averaged received power $P(k,\theta)$, $k \in \{1,..9\}$ and $\theta \in \Theta$ is given by:

$$P(\theta,k) = \frac{1}{B}\sum_{b=1}^{B} \log\left(\sum_{s=k}^{k+2} \left|\mathbf{a}^{*T}(\theta)\mathbf{h}_{s,b}^{RX,0}\right|^2\right) \tag{13}$$

The azimuth direction θ_0 and the delay tap index k_0 which correspond to the maximum received power from the desired user is consequently expressed as

$$(\theta_0, k_0) = \arg_{\theta\in\Theta, k\in[1,...,9]} \max P(\theta,k) \tag{14}$$

The value of B busts determines the averaging period. In [Mog97a] it was suggested to use B=21. This corresponds to an averaging window of approx. 100 ms, when uplink DTX is deactivated and approx. 520 ms when uplink DTX is active. For slow moving or stationary mobile stations, the time averaging over B bursts does not ensure any averaging over fast fading unless frequency hopping (FH) is being used. Random FH combined with the logarithmic power averaging over B bursts has the additional advantage of reducing the impact from strong co-channel interfering users.

The above described DoA estimation algorithm has been tested for both modelled and measured 2-D radio channel data. An Example, for a measurement route in Aalborg is given in Figure 14. The figure shows both the DoA estimate with DTX off (a) and DTX active (b). It should be noticed that frequency hopping has not been applied in the results shown although it would be an advantage.

It can be observed from Figure 14a that the DoA algorithm occasionally estimates a direction quite different from that of the mobile station. Inspection of the PAS at these locations reveals that the PAS exhibits a deep fade in the direction of the mobile station during a period, which is longer than the averaging window [Mog97]. Results on DoA estimation reported in [Tan97] and [And97] also show occasionally large excursions from the 'correct' azimuth direction. These results indicate that FH is important for a DoA algorithm to work properly in environments with a large azimuth spread (in a combination with low speed users). The similar results for a longer time averaging window, which corresponds to uplink DTX being active, is shown in Figure 14b. The excursions in azimuth have now been removed. It should be emphasised that the designed DoA algorithm is not intended to estimate the *true* azimuth direction of the mobile station, but designed to find the azimuth direction, where the strongest signal paths averaged over fast fading are received. C/I performance for the designed beamsteering algorithm is show in section 8.3.

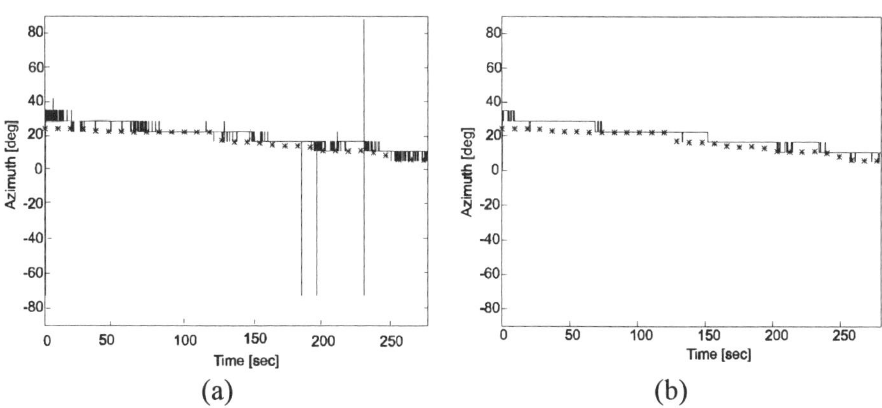

Figure 14. Examples of DoA estimates for a test route in Aalborg. In the case (a) DTX is deactivated, which approx. corresponds to a 100 ms averaging window. In (b) DTX is

active, which corresponds to an averaging period of approx. 520ms. The DoA computed from GPS is shown by : "*".

8.2 Reception

For uplink reception the beamsteering algorithm, which is proposed for downlink transmission, can successfully be used. But it is also possible to use a better combining scheme: *Optimum Combining*:

OPTIMUM COMBINING :

The narrowband approach of optimum combining [win84] can be effectively used in GSM for areas with low time dispersion (this corresponds to the GSM TU and RA profiles). In [win94] it is shown that using optimum combining with *K+N* uncorrelated antenna elements, it is possible to null out N-1 interfereres and provide K+1 diversity improvement against multipath fading. For the case of a uniform array with an element spacing in the order of $\lambda/2$, adjacent elements in the antenna array are highly correlated, see Figure 6. Hence, in order to efficiently null out interfering signals at low AS values, they must be separated in azimuth from the desired signal. For a uniform linear array with M=8 and an element spacing of $\lambda/2$, the outer elements in the array are de-correlated even for AS values in the order of 3-5°, and some diversity improvement can be achieved (see Figure 6).

For the case of *RCS* mode there are many interfering co-channel signals, and hence the interference suppression capabilities of optimum combining may be limited (relative to beamsteering). The situation is probably better for the *SCR* mode, where only a few co-channel signals are present, which are widely separated in azimuth (see [Rat96]).

The general structure of the *narrowband* optimum combining beamformer is shown in Figure 15.

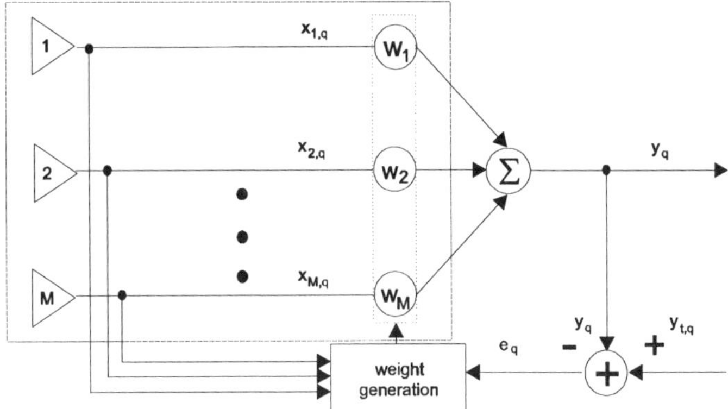

Figure 15: Optimum combining structure for an M element antenna array with controllable weights.

The goal of optimum combining is to adjust the antenna weights w_i in a Minimum Mean Square Error (MMSE) sense, such that the combined signal y_q is as close as possible to the training sequence $y_{t,q}$. The error signal e_q is the difference between the received and the a-priori known signals for symbol q.

$$e_q = y_{t,q} - y_q = y_{t,q} - \mathbf{w}^T \cdot \mathbf{x}_q \tag{15}$$

The weight vector **w** can be found by either using Direct Matrix Inversion DMI or a recursive algorithm, e.g., Least Mean Square (LMS). In order to find the optimal weight vector the squared error of Eq. 15 is calculated:

$$E\left[e_q^2\right] = E\left[y_{t,q}^2\right] - 2\mathbf{w}^T \cdot \mathbf{r}_{x,y_t} + \mathbf{w}^T \mathbf{R}_{xx} \mathbf{w} \tag{16}$$

where $\mathbf{R}_{XX}$ is the spatio-temporal correlation matrix of the received signal, and $\mathbf{r}_{x,y_t}$ is the cross-correlation vector between the training sequence y_t and the received signal x. The optimal weight solution for DMI is found by taking the gradient of Eq. 16 with respect to the weight vector **w.** Then the well known Wiener solution is then obtained

$$\mathbf{w} = \mathbf{R}_{xx}^{-1} \mathbf{r}_{x,y_t} \tag{17}$$

where

$$\mathbf{R}_{xx} = \frac{1}{Ns} \cdot \sum_{q=1}^{Ns} \mathbf{x}_q \cdot \mathbf{x}_q^{*T} \tag{18}$$

and ' $*T$ ' denotes conjugate transpose and Ns denotes the number of samples in the training sequence.

The received spatial data vector for each symbol q has the following appearance

$$\mathbf{x}_q = \left[x_{1,q} \cdot x_{2,q} \ldots\ldots x_{i,q} \ldots. x_{M,q}\right]^T \tag{19}$$

Similarly to $\mathbf{R}_{XX}$, $\mathbf{r}_{x,y_t}$ is computed as

$$\mathbf{r}_{x,y_t} = \frac{1}{Ns} \cdot \sum_{q=1}^{Ns} \mathbf{x}_q \cdot y_{t,q}^* \tag{20}$$

In GSM the antenna weights for each burst, b, can be calculated as

$$\mathbf{w}_b = \left(\sum_{q \in Whole\,burst} \mathbf{x}_{q,b} \cdot \mathbf{x}_{q,b}^{*T} \right)^{-1} \cdot \left(\sum_{q=q_0}^{q_0+25} \mathbf{x}_{q,b} \cdot I_{q-q_0,b}^{0,*} \right) \tag{21}$$

where $\mathbf{x}_{q,b}$ is the received spatial data vector for symbol q in burst b [Mog97a]. The first term in Eq. 21 corresponds to the spatio-temporal correlation matrix of the received data vector $\mathbf{x}_q$ averaged over the whole burst b. The second term in Eq.(21) corresponds to the correlation between the received data vector $\mathbf{x}_q$, and the MSK modulated training sequence I^0 of the desired user. The training sequence I^0 consists of 26 symbols. Furthermore q_0 is the estimated time instant of the received signal, which corresponds to the start of the training sequence. The time reference $q_0 \in [0, 10]$ is found as:

$$q_0 = \arg_{q_0} \max \sum_{b=bn-20}^{bn} \log\left(\sum_{q=q_0}^{q_0+25} \left\| \mathbf{x}_{q,b} \cdot I_{q-q_0,b}^{0,*} \right\|^2 \right) \tag{22}$$

where bn is the start burst number of the time averaging window. In the findings of q_0 it has been proposed to average over 21 bursts in order to average over short term fading and reduce the impact from strong cochannel interfering signals [Mog97a]. Estimation of the timing instant q_0 has been found critical for the MMSE weight optimisation. Improved performance can be achieved by a parallel demodulation of the received signal applying different antenna weights $\mathbf{w}_b$, which are computed for each of the time instants: q_0-1, q_0 *and* q_0+1, and then perform a post-detection combining/selection, for example based on the *metrics* of a Viterbi Algorithm. When many interfering signals are present it has been found that a straight forward DMI implementation does not work satisfactory. Instead a DMI implementation using Singular Value Decomposition (SVD), where only the two strongest eigenvectors are used, has successfully been implemented.

In areas with large delay spread a wideband optimum combining algorithm should be used. A such wideband optimum combing algorithm employs a tapped delay-line equaliser for each antenna element. The antenna weights of the space-time equaliser structure can similarly be calculated from Eq. 17, but using the following data vector $\mathbf{x}_{q,b}$ [win94].

$$\mathbf{x}_{q,b} = \left[x_{1,q} .. x_{1,q-1} ... x_{1,q-L+1} ... x_{i,q} .. x_{i,q-1} ... x_{M,q-L+1}\right]^T \tag{23}$$

where L is the number of delay taps. The resulting antenna weight vector is of the form

$$\mathbf{w} = \left[w_{11} . \quad . \quad . \quad w_{ML}\right]^T \tag{24}$$

The structure of the of the space-time beamformer is shown in Figure 16.

Implementation of optimum combining in GSM potentially raises several difficulties: For *SCR* mode (and also *RCS* with synchronised base stations) the issue of poor cross-correlation properties of the GSM training sequences may accomplish poor weight estimation. For *SCR* mode it is, however, possible to find pairs of training sequences with good cross-correlation properties. For an asynchronous network, the inter-cell interfering bursts are not aligned with the desired user. This is an advantage in terms of cross-correlation spill-over, but the asynchronous network implies that the co-channel signals changes during a burst, which will in contrast degrade the performance.

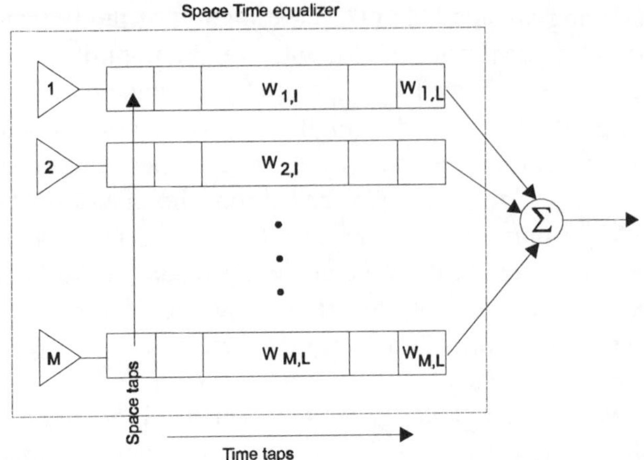

Figure 16. Diagram of a space time equaliser structure. For each antenna element there is a number of time delay taps to equalise the effects of multipath propagation. The signals from co-channel interferers are jointly reduced by the spatial beamforming for each delay tap.

8.3 Performance results

In this section the emulated C/I performance of both the beamsteering algorithm described in section 8.1 and the optimum combining algorithm described in section 8.2 are presented. The channel data for the desired user are taken from a measurement route conducted in Århus (aa_x14_00) with a base station antenna height of 32 meters (approx. 5° AS). The azimuth sweep of the test mobile ranges from +20° to 0°, and the distance to the base station varies between 1.5km and 2km. The cellular network layout shown in Figure 13 is used for emulating the C/I performance. The network uses random FH and DTX but RF-power control is not used. The frequency reuse is 1/3 and the network is 100% loaded. Both the asynchronous and synchronous network configuration has been used. Figure 17 shows the cumulative frequency of C/I for the cases of a single element, and an 8 element linear antenna array using beamsteering or optimum combining.

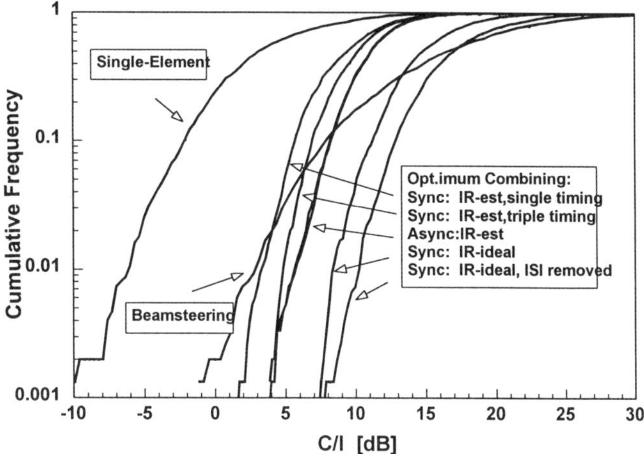

Figure 17. Cumulative frequency of C/I (averaged over 8 timeslots): For a single element, and for an 8 element linear antenna array using beamsteering or optimum combining.

It can be observed that the performance of beamsteering is close to the theoretical gain of 10log(M)≈9dB at a low outage level (c.f.<0.1)[8]. In this region the slope of the c.f.-curve for beamsteering is similar to the single element case, which implies no diversity gain. For optimum combining the curves are more steep, which implies a diversity gain. The slope of the curves are almost similar for the different options of implementation, but the mean gain differs by up to 7dB. Some interesting observations for the implementation of optimum combining in GSM can be made:

- For the case of a synchronous network the performance when only using the single timing estimate q_0 (*IR-est, single timing*) is approx. 1.5dB worse than when doing parallel optimum combining for the time instances q_{0-1}, q_0, and q_{0+1} (*IR-est, triple timing*). Nearly 4dB better performance is achieved if having an ideal channel estimate *(IR-ideal)* compared to (*IR-est, triple timing*). In [Pok97] it has been suggested to introduce new sets of training sequences in GSM with improved cross-correlation properties. A such change of training sequences will bring the performance of optimum combining in GSM closer to the case of

8 The reader may observe that at the median outage level (c.f.=0.5), the C/I gain is nearly 14 dB relative to a single element. This very high median gain value is a result of a non-uniform azimuth trajectory of the selected measurement route in a combination with the co-channel interference not being uniformly distributed in azimuth. Hence, it is not a general result.

IR-ideal. For reference purpose, the example of a flat fading channel is also shown in the figure (*IR-ideal, ISI removed*).
- The performance of optimum combining in an asynchronous GSM network (*Async: IR-est*) improves by approx. 1.5dB over *Sync:IR-est, triple*. For the asynchronous network, there is only little gained by using *triple timing*. It is somewhat surprising that the improved cross-correlation properties with random co-channel bit stream (instead of different training sequences), more than compensates for the loss of having dis-aligned co-channel interfering bursts.
- It should be noticed that optimum combining algorithm (*Async: IR-est*) only gives better C/I performance than beamsteering for outage thresholds of less than 10 %. At the 1% level, optimum combining gives approx. 3 dB additional gain over beamsteering.

9. GSM SYSTEM ASPECTS OF RCS

In this section we briefly discuss some system and network issues of deploying *Reduced Cluster Size* (RCS) in GSM:

NETWORK AND RR MANAGEMENT ISUES

When using *RCS* mode to exploit the capacity potential of adaptive antennas several radio resource (RR) management issues should be considered:

The use of adaptive beamforming is restricted to non-BCCH carriers, as the BCCH carrier must be broadcasted over the whole cell area. In terms of DoA estimation the cases of call-setup and hand-over have been identified as critical for the algorithm performance due to lack of time averaging [Mog97a]. At initial call-setup the mobile station transmits a RACH burst on the BCCH duplex carrier. In areas with large AS, the single RACH burst is insufficient to achieve a reliable DoA estimate to be used for downlink transmission. There are several methods to overcome this problem: By using the early assignment (EA) procedure in GSM [Mou92], and locate all SDCCH channels on the BCCH carrier (or on a carrier with 'normal' cluster size), downlink beamforming is not required during the SDCCH session. While the call is on the SDCCH channel, the DoA can be firmly estimated before a subsequent allocation onto a carrier using *RCS*. Alternatively, if allocating the call directly on a *RCS* carrier, the beamwidth may initially be broadened[9] until a firm DoA estimate is achieved. For the inter-cell hand-over case almost the same procedure can be used. In this case the mobile station transmits several random access bursts[10], before re-establishing the normal TCH traffic mode in the new cell. Simulation results in [Mog97a] show that 4 random access bursts are sufficient to obtain a good DoA estimate in a random FH network using a cluster size of 1/3.

RCS mode works well in a combination with other GSM capacity enhancing features, such as: RF-power control, random FH, DTX. In general, it was concluded in [Mog97a] that using *RCS* mode to exploit the potential capacity gain from adaptive antennas is also unproblematic from a network and radio resource management point of view.

CAPACITY ENHANCEMENT

It is generally anticipated that the minimum frequency reuse distance in a random FH GSM network (in a combination with DTX and RF-power control) is 3/9 (cluster size of 3 and 3 sectors per site, i.e. 9 frequencies per cluster). By using an adaptive antenna array with 6-8 elements (per sector), we estimate that the frequency reuse can be lowered to only 1/3. It should here be noticed that the 1/3 reuse is not applicable for the BCCH carrier, which must remain using a cluster size of 4/12. Table 2 gives a simple capacity estimate for a GSM network having 2 x 9.8MHz bandwidth. It is assumed that the gain from the spatial filtering is adequate to allow loading to the hard-block limit.

In the Table 2, *FH-regular* means cluster size 3/9, and *FH-Super* means cluster size 1/3. Without adaptive antennas it is estimated that the mean fractional loading of a 1/3 reuse is 0.3 [Vej95]. *FH-regular+super* means two layer reuse partitioning with a cluster size of 3/9 in the regular layer and 1/3 in the super layer. The reason for keeping some regular channels is motivated by handling call-setup and hand-over traffic, GSM phase 2+ packet switched mode - GPRS, and mobile stations with a *difficult* PAS. It can be found from Table 2 that the capacity enhancement of adaptive antennas is estimated to be a factor of 3 when using RCS mode.

9 For the Butler matrix implementation this can be done by transmitting on several adjacent beam ports.

10 To ensure this for asynchronous HO, the BTS may have to delay transmission of the RIL-3-RR PHYSICAL INFORMATION, see for example [Mou92]

Bandwidth: 9.8 MHz	Without antenna arrays		With antenna arrays	
Frequency reuse configuration	FH-regular	FH-Super	FH-super	FH-regular + super
RF-Channels	48	48	48	48
BCCH-Carriers	12	12	12	12
TCH- Carriers, Regular	36 (3/9)	0	0	18 (3/9)
TCH-Carriers, Super	0	36 (1/3)	36 (1/3)	18 (1/3)
Carriers/cell: BCCH+Reg.+Super	1+4 = 5	1+0+12 ≤ 13	1+0+12 = 13	1+2+6 = 9
TCH/F*	4 + 32 +0	4 + 0 + 0.30*96	0 + 0 +96	2 +16 +48
	37	33	96	66
Capacity limit	Hard	Soft	Hard	Hard
Erlang B, 2%	27	33	84	55
Capacity gain relative to FH regular	100 %	120%	311 %	203 %

- *Some overhead for common channels and SDCCH has been subtracted.*

Table 2. Potential capacity increase of the RCS mode of adaptive antennas. In the table it is assumed that the reuse factor on TCH frequency channels can be reduced from 3/9 to 1/3. The required C/I gain to allow this shall be in the order of 6 dB[11], which in practice corresponds to *M* equal to 6-8 elements per sector.

10. GSM SYSTEM ASPECTS OF SCR MODE

The system and network impact of using *SCR* mode (Same Cell Reuse) have only been briefly investigated by the authors. Anyway, the following discussion shortly explains some of the difficulties the authors have identified:

SDMA SIGNATURE

A major problem of introducing *SCR* in GSM is the need for assigning different signatures to SDMA users sharing the same physical channel (*t,f*). Eight different training sequences are specified for the GSM *normal burst*, in order to distinguish between the desired user and co-channel interfering users. Unfortunately the selection of training sequence is a part of the Base Station Identification Code, and hence, the training sequence number is

[11] It is assumed that an interference diversity gain of approx. 2.5 dB is achieved because the interference situation with antenna arrays is very similar to fractional loading [Olu95].

common for all channels in a cell (*normal bursts*). A solution, which allow assignment of different signatures (training sequence number) to SDMA users, is to have SDMA base station[12] separated into several logical cells.

DOWNLINK TRANSMISSION

For downlink transmission without nulling [Zet97], the spatial filtering gain is determined by the following factors: Antenna beamwidth, AS, and the azimuth separation of the SDMA users, see Sections 4 and 5. The antenna beamwidth is known a-priori, and the azimuth spread, AS can be roughly estimated for each cell, for example by evaluating the spatial cross-correlation properties between antenna elements, see Figure 6. A large AS increases the required azimuth separation of SDMA users and thereby reduces the capacity. For downlink the desired and the co-channel interfering SDMA signals are exposed to the same log-normal fading. Hence the BSS (Base Station Subsystem) can reasonably accurate predict the local C/I at the mobile station (averaged over short term fading and disregarding inter-cell co-channel interference). The C/I is a well described function of the AS, σ_θ and the azimuth separation $(\theta_0 - \theta_1)$ of the SDMA users: $C/I(\sigma_\theta, \theta_0 - \theta_1)$[13]. For a FH GSM network the required C/I, averaged over short term fading is approx. 9 dB. For a 24 element circular array and an AS of less than 5° two SDMA mobile stations shall be separated in azimuth by approx. 20°.

Downlink nulling of SDMA co-channel interfering signals is certainly a possibility [Zet97]. However, it can be observed from Figure 8 that downlink nulling will not provide substantial C/I improvement for AS values higher than 5°. The power angular spectrum will fill the nulls anyway. Considering the required accuracy of antenna array calibration [Joh95] and the DoA estimation, we estimate that commercial exploitation of null steering in GSM may be limited.

UPLINK COMBINING

The uplink performance of *SCR* mode in GSM is more complicated to analyse than the downlink. The reason for this is that the signal received from different mobile stations have been exposed to a different r^γ path loss, and uncorrelated log-normal fading. The *near-far* effect can be reduced by utilising uplink RF-power control. In GSM the uplink RF

12 Whether it is possible to do this without allocating several BCCH carriers has not been investigated.

13 It is here assumed that downlink power control is not activated. Otherwise the power control setting shall be included.

power control range is limited to less than 30 dB. Secondly the regulation loop is too slow to track rapid changes in the environment. For a TDMA system with reduced uplink power control capabilities such as GSM, the *near-far* problem of *SCR* mode can be solved by continuously sorting the users according to both their azimuth direction θ and their propagation path loss, *L*: $F(\theta, L)$. Assuming that it is possible to perform an ideal channel assignment (*t*, *f*) so the uplink RF-power control can fully compensate for the pathloss difference between SDMA users, then the spatial filtering gain shall *only* take care of the residual power-control error (and ensure a C/I of 9 dB). For example, if two mobile stations are assigned to the same physical channel (*t,f*), and the residual power control error is 5 dB (due to log-normal fading), then the standard deviation on C/I is $\sqrt{5^2+5^2} \approx 7\text{dB}$. For an outage probability of less than 5%, the spatial filtering gain shall be 14dB (two times the standard deviation) in order to compensate for the residual power control error. The required spatial filtering gain in uplink is therefore (14+9)dB=23dB. Several techniques may be used to obtain this high filtering gain: Optimum combining, joint-detection MLSE or/and beamsteering with a window function to suppress sidelobe level.

NETWORK AND RR MANAGEMENT ISUES

The hand-over rate in a network using *SCR* is likely to increase significantly: When a mobile station roams, the sorting according to $F(\theta, L)$ changes, and the physical channels in a cell (*t,f*) shall continuously be re-allocated by means of an intra-cell hand-over. A great advantage of *SCR* mode compared to *RCS* mode is, however, that it can be installed in a single cell to boost the capacity in a "hot-spot" area, whereas the *RCS* mode needs to be installed in a larger area of the network to be very efficient. As previously mentioned the authors have not made any extensive capacity analysis for the concept of *SCR* applied to GSM as this mode may give some difficulties with respect to training sequence allocation. The solution of splitting a SDMA base station into several logical cells combined with the $F(\theta, L)$ sorting of mobiles across the logical cells requires major changes to the BSS software.

11. CONCLUSIONS

Based on the presented measurement and simulation results we conclude that adaptive antennas is a very promising technology for both range extension and capacity enhancement of macro-cellular base stations.

Two different network strategies can be used to convert the spatial filtering gain of adaptive antennas into a capacity increase: (i) Reduced Cluster Size (RCS) and (ii) Same Cell Reuse (*SCR*). We expect that the concept of *RCS* can be directly introduced in current GSM, whereas the concept of *SCR* may give some difficulties. An estimate of the potential capacity gain for *RCS* mode is a factor of 3.

For downlink beamsteering in *RCS* mode a simple and robust direction of arrival algorithm has been designed. The same algorithm can with success also be used for uplink reception. The more complex Optimum Combining scheme has also been tested: The C/I gain for an eight element linear antenna array using beamsteering has been found to be close to the theoretical value of 9dB (AS less than 5°). The gain of optimum combining in an asynchronous network is found to be only 2-3dB higher than for beamsteering. Optimum combining in a synchronous network can potentially give better performance, if the training sequences are modified to have better cross-correlation properties as proposed in [Pok97].

It has been shown that a large azimuth spread of the radio channel has a significant impact on the spatial filtering performance of beamsteering (which is used for downlink). The azimuth domain approach of adaptive antennas may therefore not be the best solution for base station installations seeing a large angular spread (such as micro- and pico-cell installations). Distributed antenna elements may in this case be a better approach for adaptive antennas.

A very important issue of adaptive antennas, which has not been covered, is of course the cost efficiency compared to other capacity enhancing technologies. Unfortunately, the authors are incapable of answering this question.

ACKNOWLEDGEMNTS

The help of the authors' colleagues at CPK, Kim Olesen, Steen-Leth Larsen and Frank Frederiksen, who has been responsible for the hardware construction and measurements are highly appreciated. Prof. J. Bach Andersen is thanked for his careful and critical reading of the manuscript.

The authors also wish to thank the European Commission, the Danish research Council (STVF) and Nokia Telecommunication for the financial support of the presented work.

REFERENCES

Ana Anaren Microwave, Inc., "Antenna Feed Networks", *Data Sheet*

And97 Anderson, S. et al., "Ericsson/Mannesmann GSM Field-Trial with Adaptive Antennas", Proc. of EPMCC'97, Bonn 97, pp. 77-85

Bel63 P.A Bello, "Characterization of Randomly Time-Variant Linear Channels", IEEE Trans. on Communication Systems, Vol. 11, pp. 360-393, December 1963.

Bot95 Bottomley, G.E. and K. Jamal," Adaptive arrays and MLSE equalization", IEEE Proc. of VTC'95, Chicago, pp. 50-54.

Egg95 Eggers, Patrick, "Angular Dispersive Mobile Radio Environments Sensed by Highly Directive Base Station Antennas", IEEE Proc. Personal Indoor and Mobile Radio Communications (PIMRC'95), pp. 522-526, September 1995.

For94 Forssén, Ulf, J. Karlsson, B. Johannisson, M. Almgren, F. Lotse, and F. Kronestedt, "Adaptive Antenna Arrays for GSM900/DCS1800", IEEE Proc. of VTC '94, Stockholm, Sweden, pp. 605-609.

Jak74 W.C. Jakes, "Microwave Mobile Communications", IEEE Press, Piscataway, New Jersey, USA, 1974.

Joh95 Johannison, Bjørn, "Planar Antenna Array for Adaptive Beamforming". Proc. of Nordic Radio Symposium 1995, pp. 177-180, Saltsjobaden, Sweden, April 1995.

Joh97 Johannison, B., "Active and adaptive base station antennas for mobile communication", Proc. of Antenna '97, Stockholm, Sweden.

Lee73 Lee: W., "Effects on Correlation Between Two Mobile Radio Base-Station Antennas", IEEE Trans. on Communications, Vol. 21, No. 11, November 1973.

Meh94 Mehrota, Asha, "Cellular Radio: Analog and Digital Systems", Artech House, Inc. 1994, ISBN:0-89006-731-7.

Mog90 Mogensen, Preben E., "Proposal for 1800 MHz Hata Alike Model", COST231 TD(90)-115, Darmstadt, Dec. 1990.

Mog91 Mogensen, P.E., C. Jensen and J.B. Andersen, "1800 MHz mobile netplanning based on 900 MHz measurements", COST231 TD(91)-008, Firenze, Jan. 1991.

Mog96 Mogensen, P.E., F. Frederiksen, H. Dam, K. Olesen, and S.L. Larsen, "A hardware testbed for evaluation of adaptive antennas in GSM/UMTS", IEEE Proc. of PIMRC '96., Taipei, Taiwan, pp. 540-544

Mog96b Mogensen, Preben E. and Jeroen Wigard, "On Antenna and Frequency Diversity in GSM Related Systems (GSM900,DCS1800,PCS1900)", IEEE International Symposium on Personal Indoor and Mobile Communications, Taipei, Taiwan, October 1996, pp. 1272-1276

Mog97a Mogensen, P.E., P. Zetterberg,H. Dam,P.L.Espensen, F. Frederiksen, "Algorithms and Antenna Array Recommendations", technical report AC020/AUC/A1.2/DR/P/005/b1, 1997

Mog97 Mogensen, P.E., K.I. Pedersen, P. Leth-Espensen, et al. "Preliminary Measurement Results From an Adaptive Antenna Array Testbed for GSM/UMTS", IEEE Proc. of

VTC'97, pp. 1592-1596, May 1997.

Mou92 Mouly, Michel and Marie-Bernadette Pautet, "The GSM System for Mobile Communications", Michel Mouly and Marie-Bernadette Pautet, 49 rue Loise Bruneau5, F-91120 Palaiseau France, 1992. ISBN 2-9507190-0-7.

Nor94 Nørklit, O. and J.B.Andersen, "Mobile Radio Environments and adaptive arrays", IEEE Proc. of PIMRC'94, Hague, Sept. 18-22 1994, pp.725-728.

Olu95 Olufsson, Håkon, Jonas Näslund, Benny Ritzén, and Johan sköld, "Interference diversity as means for increased capacity in GSM", Proc. of EPMCC '95, Italy Nov. 1995, pp. 97-102

Ped97a Pedersen, K.I., P.E. Mogensen and B.H. Fleury, "Power azimuth spectrum in outdoor environments", IEE Electronics Letters, Aug. 28 1997, Vol. 33, No. 18, pp 1583-1584

Ped97b Pedersen, K.I., P.E. Mogensen and B.H. Fleury, " Analysis of Time, Azimuth and Doppler Dispersion in Outdoor Radio Channels", Proc. of ACTS Mobile Summit '97, Aalborg October 1997, pp. 308-313.

Pok97 Pokkila, M., P.Ranta, "Channel Estimator for Multiple Co-channel Demodulation in TDMA Mobile Systems", To appear in Proc. of EMPCC'97, Bonn, Sept. 1997.

Rat96 Ratnavel, S., A. Paulraj, A.G. Constantinides, "MMSE Space-Time Equalization for GSM cellular Systems, IEEE Proc. of VTC '96, Atlanta, April 1996, pp.331-335.

Sal93 J. Salz, J. Winters, "Effect of Fading Correlation on Adaptive Arrays in Digital Wireless Communications", IEEE Proc. ICC'93, Geneva, Switcherland, pp. 1768-1774, May 1993.

Tan95 Tangemann, M., M. Beach et al., "Report on the Benefits of Adaptive Antennas for Cell Architectures", RACE TSUNAMI R2108/SEL/WP3-4/DS/P/029/b1, May. 1995

Tan97 Tangemann, M., U.Bigalk, C. Hoek, M. Hother, "Sensitivity Enhancements of GSM/DCS 1800 with Smart Antennas, Proc. of EPMCC'97, Bonn 97, pp. 87-94

She61 J.P. Shelton, K.S. Kelleher, "Multiple Beams from Linear Arrays", IRE Trans. on Antennas and Propagation, pp. 151-161, March 1961.

Veen88 B.D. Veen, K.M. Buckley, "Beamforming: a versatile approach to spatial Filtering", IEEE ASSP Magazine, pp. 4-24, April 1988.

Vej95 Vejlgaard, Benny and Jesper Johansen, "Capacity analysis of a Frequency Hopping GSM System", Master Thesis Report, Aalborg University, June 1995

Win84 Winters, Jack H, "Optimum Combining in Digital Mobile Radio with Co-channel Interference", IEEE Journal on Selected Areas in Communications, Vol. SAC-2, No. 4, July 1984.

Win94 Winters, J.H., "The impact of antenna diversity on the capacity of wireless communications systems", IEEE Trans. on Communications, Vol. 42, No. 2/3/4, Feb/Mar/Apr. 1994, pp. 1740-1751.

Zet97 Zetterberg, P., "Mobile Cellular Communications with Base Station Antenna Arrays: Spectrum Efficiency, Algorithms and Propagation Models", Ph.D. thesis 1997, KTH, Stockholm, ISSN 1103-8039

Chapter 2

Interference suppression by joint demodulation of cochannel signals

PEKKA A. RANTA, MARKKU PUKKILA
Nokia Research Center

Key words: Joint detection, channel estimation, training sequences, interference

Abstract: Inter-cell cochannel interference (ICCI) is an inherent problem in all cellular systems due the necessity to reuse the same frequencies after a certain reuse distance. In GSM, the fact that the number of nearby cochannel interferers is relatively small leads to a high probability of a dominant interferer (DI). Hence, suppression of DI alone provides substantial capacity improvement for GSM. The paper summarises different aspects of interference suppression by joint demodulation of cochannel signals in the GSM system. The probability of DI is investigated by network simulations. Moreover, receiver algorithms are described and receiver performance analysis is provided. In addition, requirements that the application of the technique poses for the GSM systems are explained.

1. INTRODUCTION

Availability of the radio spectrum will be one of the main concerns in the future mobile radio systems as the number of mobile users is rapidly increasing and new data services are taken into use. In GSM, one of the most important factors limiting the cellular capacity is the cochannel interference (CCI) originating from the surrounding cells using the same carrier frequencies. Current GSM has already introduced a number of advanced radio access techniques such Discontinuous Transmission (DTx) and power control (PC) to minimise the problem of cochannel interference.

In addition, GSM supports slow frequency hopping (FH) to overcome fast fading and provide interference diversity. [Mou92]

To improve the capacity of GSM even further it is a natural choice to take advantage of rapid development of digital signal processing techniques and consider cochannel interference (CCI) suppression techniques implemented in receivers. The receivers' improved susceptibility to CCI provides means for lowering the frequency reuse distance in the network, that is, increase in the network spectrum efficiency. Alternatively, data rates can be improved by reducing channel coding or just take the gain to increase quality of service. For example, in GSM packet radio (GPRS) and Adaptive Multirate Codec (AMR) several channel coding rates are suggested allowing an individual terminal to benefit from higher data rate whenever it is possible [ETSa,ETSb]. Another advantage of the IC-receiver is that it eases frequency planning as the system becomes more robust against interference. This aspect may be important especially when implementing high capacity low reuse cellular systems.

Interference cancellation techniques applicable in GSM can be divided into three categories: interference cancellation by joint (or multiuser) detection of cochannel signals, blind or semi-blind methods and adaptive antennas. The first category of interest in this paper has been earlier studied, e.g., in [Gir93, Yos94, Wal95, Ran95a, Ran95b, Ran96, Edw96, Ran97a, Ran97b]. Most of the papers concentrate on the receiver techniques, but in [Ran95b, Ran96] capacity estimates are given showing potential capacity gain up to 60% in macrocells. A blind approach for GSM CCI reduction is presented in [Ber96] using the knowledge of the constant envelope property of GMSK modulated signals. Another blind method has been introduced in [Ant97] based on usage of Hidden Markov Models. Application of neural networks for CCI cancellation is proposed in several papers, e.g., [Che92, Che94, How93]. The last category of IC-techniques is probably the most powerful against interference and is based on the digital antenna array processing techniques with interference rejection combining (IRC) in the receiver [Bot95, Win84, Fal93, Karl96, Esc97, Ran97c]. However, digital antenna array processing techniques require multiple RF receivers for which reason they are primarily applicable in the base station receivers. The interference cancellation by joint detection (JD/IC) requires only a single antenna receiver making it an attractive alternative especially in the mobile receivers.

In the conventional GSM receivers CCI is treated as additive Gaussian noise. The fact that CCI is deterministic in nature and partly known, e.g., modulation type and possible training sequence codes, makes multiuser detection (MUD) or joint detection (JD) techniques feasible in GSM receivers. In CDMA systems, MUD techniques are well-investigated for the rejection of intra-cell interference which is the primary source of cochannel interference in CDMA systems [Mos96]. In GSM, as the users are orthogonal within a cell, the problem is purely to combat inter-cell cochannel interference. In this case the number of cochannel signals is much fewer and often a dominant interference (DI) exists which allows to reduce the baseband receiver complexity with only a reasonable performance loss.

In this paper different aspects of application of the JD/IC technique in the GSM network will be summarised. First the problem of cochannel interference is introduced in more detail and it is shown that the probability of DI can be relatively high in GSM networks. The receiver algorithm is described based on the earlier contributions including detection algorithm and channel estimator with DI identification. The receiver complexity is discussed and complexity reduction methods are suggested. The performance of the technique is evaluated using a novel link simulator introduced in [Ran97b] using interference distribution information from a network simulator. The requirements of the JD/IC receiver from the systems point of view are considered in detail and the potential applications for the technique are enlightened. Finally, conclusions are drawn.

2. COCHANNEL INTERFERENCE

In mobile networks, the desire to maximise the number of available traffic channels in a given geographical area results to cochannel interference (CCI) and adjacent channel interference (ACI) problems. In this paper, we are interested in the removal of CCI only although in principle the removal of ACI could be possible as well.

2.1 Frequency reuse

Cellular systems exploit the concept of frequency reuse meaning that the same frequencies are repeated according to a certain *reuse pattern* or *reuse distance*. To maximise the network capacity [users/MHz/cell] we wish to minimise the number of cells in the reuse pattern. However, frequency re-usage causes inherent cochannel interference (CCI) problems in receivers. Hence, reuse pattern cannot be reduced without loss in the quality of service. Evidently, cellular capacity can be improved if receivers' susceptibility to the interference can be enhanced.

Fig. 1 below illustrates a co-channel communications situation in an idealised cellular network with hexagonal cells. In this case, the problem is described in the downlink direction, that is, the mobile (MS) is the receiving end. In GSM, users are orthogonal within a cell, thereby the cochannel interference purely originates from the surrounding cells and the number of CCI sources is rather low, i.e. six in this case. In a real network, more interferers further off from the centre cell exist but they contribute less to the total interference as the distance is increasing.

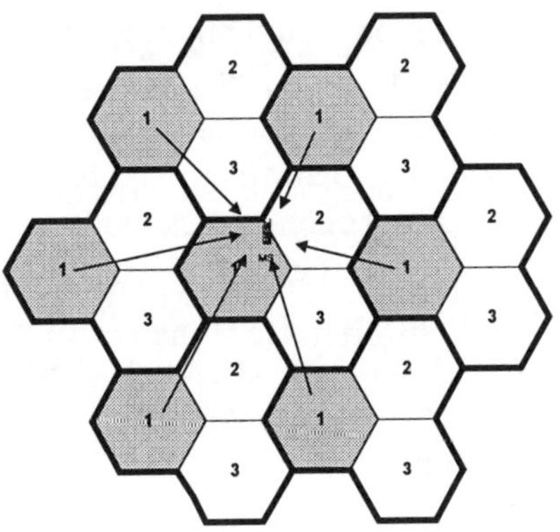

Figure 1. Cochannel interference problem in GSM. Downlink direction and reuse pattern three.

In Fig. 1, signal to noise and interference ratio (SNIR) experienced by the receiver can be described by

$$SNIR = \frac{C}{\sum_{i=1}^{N} I_i + N_{RX}} \tag{1}$$

where C is the desired signal power, I_i is the power of an incident cochannel signal and N_{RX} is the receiver noise power. The cochannel signals $\{I_i\}$ propagate through independent channels undergoing multiplicative effects of lognormal shadowing, Rayleigh fading, distance attenuation and power control, and therefore they probably differ very much in their power levels. Still, it is likely that lognormal shadowing, which is caused by obstacles on the propagation path, will dominate the distribution due to its long distribution tails. Hence, the interference observed by a single receiver is a sum of lognormally distributed interfering signals. The analysis in [Bea96, Stu96] suggests that the sum of lognormal signals is still close to the lognormal distribution. Hence, with frequency hopping the interference level in each independent hop may be approximated by lognormal distribution.

2.2 Existence of dominant interferer

In GSM, a dominant interferer (DI) likely exists since the number of nearby cochannel interferers is rather small, for example, in the case of omnidirectional cells the nearest cochannel tier includes six cells (see Fig. 1). This number is further limited to two or three by cell sectorisation or usage of adaptive antennas. In addition, discontinuous transmission (DTx) as well as fractional loading cause that the interferers do not likely transmit simultaneously. Furthermore, independent distance attenuation, shadowing, Rayleigh fading and power control make the power levels of the received signals probably very much different from each other.

Obviously, cancellation of DI alone can improve the receiver performance significantly with the advantage of remarkably lower complexity than suppressing more interfering signals. When channel intersymbol interference (ISI) is moderate, it is also practical to consider joint demodulation of more than two cochannel signals. The efficiency of the DI cancellation is naturally dependent on the dominant to rest of interference ratio (DIR) in addition to the signal to noise and interference ratio (SNIR). In mathematical terms DIR can be expressed as

$$DIR = \frac{I_{dom}}{\sum_{i=1}^{K} I_i - I_{dom} + N_{RX}} \tag{2}$$

where I_{dom} is the dominant among all the interfering signals, i.e., $I_{dom} = \max(I_1, I_2, ..., I_N)$ and N the receiver noise power. In the further analysis it is assumed that noise power is negligible compared to the rest of interference.

In Fig 2. simulation results of uplink DIR distributions are plotted in a urban cellular network with hexagonal omnicells and reuse three. The main simulation parameters are given in Table 1 below. In the simulations Rayleigh fading is neglected as it was found to have only a minor

Table 1. Parameter values used in the simulation of a GSM mobile network.

reuse	3
number of interferers	18
duplex direction	uplink
propagation index	4
std of lognormal shadowing	8 dB
power control dynamicity	30 dB
power control error	lognormal with std 5 dB
handover margin	3 dB
base station activity	50%

improvement on the DIR distribution. The upper subplots represent the probability density function (pdf) and cumulative probability density function (cdf) of DIR measured for all mobiles, respectively, and the lower subplots represent DIR measured only for those mobiles experiencing bad quality, i.e. *C/I* below 9 dB. In the first case, we find that DIR is above 5 dB with probability of 30 %, but in the latter case the probability has been increased to 60%. Since DIR of 5 dB corresponds approximately to 3 dB IC-gain [Ran95a], we can conclude that 60 % mobiles experiencing bad quality

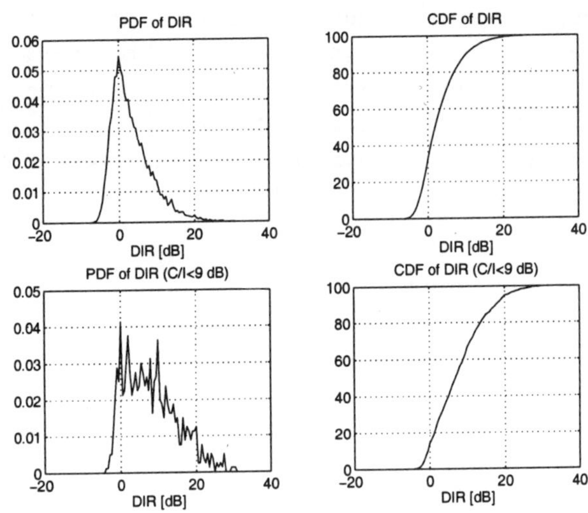

Figure 2. DIR distributions (pdf and cdf) for all mobiles (up) and for mobiles with C/I<9dB (down).

can achieve more than 3 dB IC-gain. Note that the DIR distribution would become even more favourable in case of sectorised cells.

In addition to chosen radio access techniques (DTx, PC, etc.), the DIR ratio depends on the environmental parameters such as the propagation index and shadow parameter. In practical network planning, different long term propagation characteristics for each of the cochannel signals and irregularly placed cell sites often leads to the presence of a dominant interfering signal. In microcellular environment, street crossings are known to be the most difficult places from the interference point of view. In [Ran97a] it is shown that the DI cancellation is a very powerful method to solve the interference problem in microcell street crossings.

3. RECEIVER ALGORITHMS

Fig. 3 above depicts the cochannel communications system model considered in this section. It consist of N transmitters with independent time varying channels, additive white Gaussian noise source, receiver filter, joint channel estimator (JCE) and joint detector (JD). As shown by the figure, joint detector (JD) employs directly the channel estimates provided by the joint channel estimator (JCE). The joint detector can provide, although it is not necessary, symbol estimates of the each cochannel bit stream in the process. In case of DI cancellation there are only two signals in the process i.e. N=2. In the following sections we explain the JCE and JD blocks in detail, but we start explaining the format of the received signal that is necessary for understanding the basis of joint signal detection.

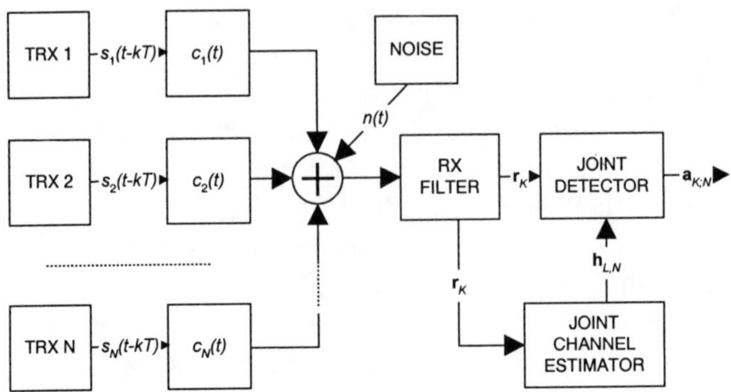

Figure 3. Cochannel communications system and receiver model

3.1 Signal format

3.1.1 Burst structure

In Fig. 4. the GSM TCH burst is described containing three zero symbol tail bits in the beginning and end, training sequence of 26 bits in the middle and two data blocks of 58 bits around the training sequence. One bit in the both sides of training sequence (two bits in total) is reserved for signalling purposes. The tail bits are used by the demodulator to initialise and finish the detection process. The training sequence is used for channel estimation enabling signal equalisation and coherent detection. [ETSc]

The training sequence consists of a reference sequence of length 16 bits, five guard symbols in the both sides of the reference sequence or, equivalently, ten guard bits before the reference sequence. The purpose of the guard bits is to cover the time of intersymbol interference and time synchronisation errors. In GSM, eight distinct training sequences are specified from which four sequences have seven zeros in their periodic autocorrelation function in both sides of the main peak and the other four sequences have six zeros around the main autocorrelation peak. This property enables estimation of at least five channel taps just based on the strict correlation of the known and received training sequence. [ETSc]

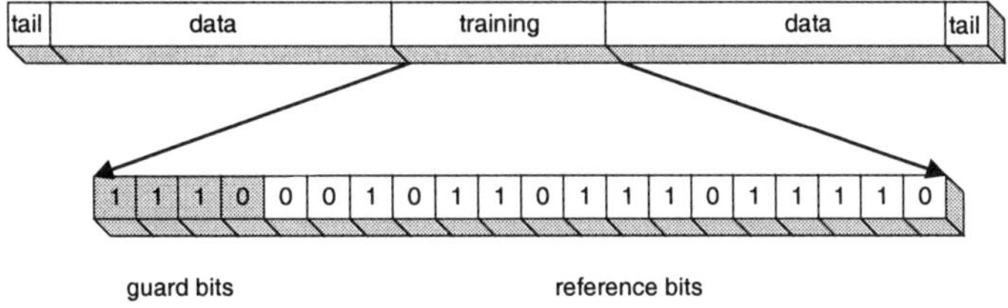

Figure 4. An example of GSM burst

3.1.2 Modulation and channel

GSM system employs non-linear GMSK modulation which is a constant envelope modulation and thereby does not pose as tight requirement power amplifier linearity as its linear PSK and QAM correspondents. Nevertheless, GMSK can be approximated as a linear modulation method since the modulator pulse can be constructed as a sum of finite number of amplitude modulated pulses [Lau86]. From this point of view GMSK can been seen as Binary Offset QAM, i.e. Offset QPSK, with a different pulse shape. In the following presentation we use this approximation of GMSK and take it as a linear modulation method with binary modulation alphabets {-1,+1}. Although the presentation is limited to binary transmission, the same equations can be easily extended to higher level modulations.

The transmitted waveform can be expressed in the complex baseband form as

$$s(t) = \sum_{k=1}^{K} a_k g(t - kT) \tag{1}$$

where $a_k \in \{-1,+1\}$ and *g(t)* is the transmitted pulse shape, *T* is the symbol period and *K* is the transmitted sequence length. The signal passes through a time-variant radio channel with the impulse response $c(t,\tau)$

$$c(t,\tau) = \sum_{i=-\infty}^{\infty} p_i \delta(t-\tau_i) \tag{2}$$

where τ_i is the delay and the complex variable p_i stands for the amplitude of the *i*th discrete multipath component in the impulse response. $\delta(t)$ is the Dirac function. In GSM, the channel can be assumed to be constant over a transmission burst because it is short enough compared to the channel fading rate in practical vehicle speeds. Hence, the channel impulse response during a burst depending only on τ can be expressed as $c(\tau)$. Often this type of channel is characterised by the name *block fading channel.*

In the cochannel communications system under consideration the received signal is a superposition of N cochannel signals present in the receiver input as described in Fig. 3. More precisely, the received signal during a burst can be expressed as

$$r(t) = \sum_{i=1}^{N} \sum_{k=1}^{K} a_k^{(i)} h^{(i)}\left(t-kT\right) + n(t) \tag{3}$$

where $h^{(i)}(t)$ is the channel impulse response in the *i*th cochannel including effects of transmitted pulse shape, radio channel, receiver filter matched to the transmitted pulse shape and noise whitening filter. The complex variable *n(t)* stands for white Gaussian noise process with two-sided power spectral density N_0. Samples obtained from $r(t)$ after the symbol rate sampler form a set of sufficient statistics for the detection of the transmitted symbols.

For channels of limited length (L taps), the sum in Eq. (3) can be written as

$$y_k = \sum_{i=1}^{N} \sum_{l=0}^{L} a_{k-l}^{(i)} h_l^{(i)} \tag{4}$$

taking $2^{N*(L+1)}$ discrete values during a transmission. The equation presents the noiseless channel output and will used in the following to describe the receiver algorithm.

3.2 Joint detection

3.2.1 Joint MLSE Detection

An optimum demodulator which *minimises the probability of sequence errors* in the presence of intersymbol interference (ISI) and white Gaussian noise is the Maximum Likelihood Sequence Estimation (MLSE) that can be implemented using the Viterbi algorithm [For72, Ung74]. For the purpose of simultaneous demodulation of multiple signals, W. van Etten [Ett76] extended Forney´s and Ungerboeck's algorithms with a specific problem of cross-talk in cable transmission systems and cross-polarisation interference in radio link transmission systems in mind. Van Etten pointed out that also in the multiple signal detection with ISI, the process at the channel output can be characterised by finite-state discrete time Markov process in memoryless noise, so both ISI and CCI expand the number of states in the Markov process. However, CCI increases the number of possible transitions between states as well, therefore it can be better characterised by the expansion of the modulation alphabet than increment of ISI.

To explain the Joint MLSE algorithm in more detail we express the problem in mathematical terms. The maximum likelihood sequence estimator would estimate the most probably transmitted sequences $\mathbf{a}_K \hat{=} (\mathbf{a}_K^{(1)}, \mathbf{a}_K^{(2)}, \ldots, \mathbf{a}_K^{(N)})$ in all N cochannels jointly from the received signal vector $\mathbf{r}_K \hat{=} (r_1, r_2, \ldots, r_K)$. Note that the known tail symbols at the end of the transmitted sequence are included in the definition of K. Hence the maximisation criterion for JMLSE becomes

$$\begin{aligned} \hat{\mathbf{a}}_K &= \arg\max_{\mathbf{a}_K} \left[p\left(\mathbf{r}_K \middle| \mathbf{a}_K^{(1)}, \mathbf{a}_K^{(2)}, \ldots, \mathbf{a}_K^{(N)}\right)\right] \\ &= \arg\max_{\mathbf{x}_K} \left[p\left(\mathbf{r}_K \middle| \mathbf{x}_{K},\right)\right] \end{aligned} \tag{5}$$

where the vector $\mathbf{x}_K \hat{=} (x_1, x_2, \ldots, x_K)$ represents the corresponding state sequence. Assuming the received signal $\mathbf{r}_K$ to be disturbed by additive white Gaussian noise samples it is convenient to use the equivalent loglikehood form and express the problem as

$$\hat{\mathbf{a}}_K = \arg\min_{\mathbf{a}_K} \left[\sum_{k=1}^{K} \left| r_k - y_k \right|^2 \right] \tag{6}$$

where y_k is defined by Eq. (4). The equation returns the minimum sum of Euclidean distances over all possible sequences. It is well known that this minimisation problem can be solved by the Viterbi algorithm using the recursion

$$J_k(\mathbf{a}_k^{(n)}) = J_{k-1}(\mathbf{a}_{k-1}^{(n)}) + |r_k - y_k|^2 \tag{7}$$

where the term $J_{k-1}(\mathbf{a}_{k-1}^{(n)}),\ n = 1,2,\ldots,N$ presents the survivor path metric at the previous stage in the trellis. In fact, the path metrics of the single signal detection is identical to Eq. (7) using y_k in Eq. (4) with N equal to 1. In other words, the difference is that in every symbol period JMLSE weights the symbols $(a_k^{(1)}, a_k^{(2)}, \ldots, a_k^{(N)})$ jointly instead of $a_k^{(1)}$ alone.

3.2.2 Joint symbol-by-symbol MAP detection

In difference to MLSE detection minimising the sequence error probability, the objective of the Maximum a Posteriori (MAP) algorithm is to *minimise the probability of a single symbol error* by estimating a Posteriori probabilities (APP) of states and transitions of the Markov source from the received signal sequence. Type-I MAP algorithm introduced by Chang and Hancock [Cha66] uses the information of the whole received sequence to estimate a single symbol probability. In other words, type-I MAP algorithm selects the symbol $a_k^{(i)} \in \{-1,+1\}$ at time instant k which maximises the following APP

$$\hat{a}_k^{(i)} = \arg\max_{a_k^{(i)}} \left[p\left(a_k^{(i)} \middle| \mathbf{r}_K\right) \right] \tag{8}$$

where $\hat{a}_k^{(i)}$ is the symbol estimate in the cochannel i of interest. Type-II MAP algorithm introduced by Abend and Fritchman [Abe70] instead uses all the information from the previous samples but only a limited amount of the future samples determined by a fixed lag decision delay. MAP type-II algorithm maximises the following a posteriori probability

$$\hat{a}_k^{(i)} = \arg\max_{a_k^{(i)}} \left[p\left(a_k^{(i)} \middle| \mathbf{r}_{k+D}\right) \right] \tag{9}$$

where $\mathbf{r}_{k+D} \mathrel{\hat{=}} (r_k, r_{k+1}, \ldots, r_{k+D})$ and D stands for the decision delay. When the decision delay is increased, the performance of the type-II MAP algorithm approaches to that of the type-I MAP algorithm [Li95].

The MAP algorithms require multiplicative accumulation of the transition probabilities which is not desirable in ASIC implementation while in DSP implementation the cost of multiplication is much less significant. Nevertheless, this problem can be avoided with only a minor performance loss by computing the metric of the MAP algorithm in log-domain leading to additive accumulation of path metrics [Li95, Rob95]. The complexity of these algorithms approaches to that of Viterbi algorithm. The main advantage of MAP algorithms over the Viterbi algorithm is more reliable transfer of soft information for the channel decoder.

3.3 Joint channel estimation

The estimation of the channel impulse response for both desired and dominant interfering signals is a crucial matter for joint detection. To be able to compute y_k in Eq. (7) or the probabilities in (8) and (9), knowledge of the channel impulse response of all jointly demodulated cochannels is required. In the conventional GSM receiver, channel estimation is based on a priori known training sequences as explained in Sec. 3.1. Evidently, JCE can also exploit the knowledge of training sequences carried by cochannel signals. However, the cross-channel interference between cochannel signals makes the task very challenging.

A most straightforward method to solve the channel estimates fastly, reliably and accurately is to use a one-shot channel estimation based on a solution of a system of linear equations [Ste94, Ran95a, Puk97] posing two additional constraints for the system:

1. Training sequences in each cochannel should be received simultaneously (Fig. 5) to be able to remove the cross-channel interference.
2. Training sequences should be unique with good cross-correlation properties at least in the closest cochannels.

The first requirement is not fully strict in the sense that some asynchronism can be allowed depending on the reference and guard period lengths of the training sequence. The second requirement implies some sort of code sequence planning and selection of best training sequences among the existing ones or totally new sequence sets for GSM. These requirements are more thoroughly discussed in Sec. 5.

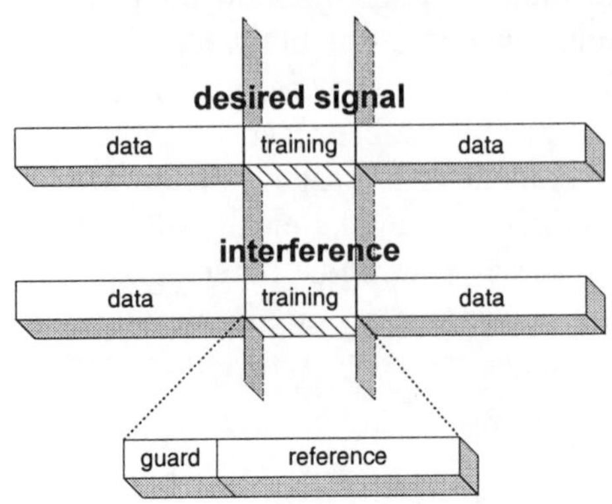

Figure 5. Two cochannel bursts received simultaneously

Asynchronous system

To perform joint channel estimation in the asynchronous system it is necessary to use semi-blind channel estimation methods which means that the available training sequence information is also utilised. In this case it would be beneficial to maximise the training sequence length to get a better protection for cross-channel interference. In any case, the performance in the asynchronous case will be worse than in the synchronous mode not only because of the blind channel estimation but since the interference may change during a transmission burst.

3.3.1 Basic algorithm

Suppose there are N synchronous co-channels, i.e., the primary user and N-1 interferers each having a different training sequence and a different channel. Denote the N radio channels by

$$\mathbf{h}_{L,n} \triangleq (h_{0,n}, h_{1,n}, \ldots, h_{L,n})^T, \quad n = 1,2,\ldots,N,$$

each of length (L+1) with complex channel tap weights. Collect the channel impulse responses into the vector $\mathbf{h}$ as follows

$$\mathbf{h} \triangleq (\mathbf{h}_{L,1}^T, \mathbf{h}_{L,2}^T, \ldots, \mathbf{h}_{L,N}^T)^T.$$

The number of parameters above is thus $N \times (L+1)$. The training sequence of the *n*th channel consisting of the guard and reference sequence bits is denoted by

$$\mathbf{m}_n \triangleq (m_{0,n}, m_{1,n}, \ldots, m_{P+L-1,n})^T, \quad n = 1,2,\ldots N,$$

with $L+P$ elements $m_{p,n} \in \{-1,\ 1\}$, where L is the number of the guard bits implying that maximum of $L+1$ taps can be estimated and P is the length of the reference sequence.

The received signal corresponding to the reference bits is then

$$\mathbf{y} = \mathbf{M}\mathbf{h} + \mathbf{n} \tag{10}$$

where $\mathbf{n}$ represents Gaussian noise samples with the covariance matrix $\mathbf{R}$, and the matrix $\mathbf{M} = (\mathbf{M}_1, \mathbf{M}_2, ..., \mathbf{M}_N)$ includes the transmitted training sequences organised to the matrices $\mathbf{M}_n$, n=1,2...,N, as follows:

$$\mathbf{M}_n \triangleq \begin{bmatrix} m_{L,n} & \cdots & m_{1,n} & m_{0,n} \\ m_{L+1,n} & \cdots & m_{2,n} & m_{1,n} \\ \vdots & & \vdots & \vdots \\ m_{P+L-1,n} & \cdots & m_{P,n} & m_{P-1,n} \end{bmatrix}.$$

The maximum likelihood channel estimate is given by

$$\hat{\mathbf{h}}_{ML} = (\mathbf{M}^H \mathbf{R}^{-1} \mathbf{M})^{-1} \mathbf{M}^H \mathbf{R}^{-1} \mathbf{y}, \tag{11}$$

and assuming that the noise in Eq. (10) is white it reduces to

$$\hat{\mathbf{h}}_{ML} = (\mathbf{M}^H \mathbf{M})^{-1} \mathbf{M}^H \mathbf{y}. \tag{12}$$

The result is the well-known solution of Wiener-Hopf equation in matrix form. Note that the Eq. (12) is equivalent to the conventional channel estimator if N is equal to 1.

Table 2. SNR degradation of different training sequence sets.

Set	Length	Set size	SNR degr. (dB)	
			worst pair	best pair
GSM	16 bits	7	8.0	3.2
20-BIT	20 bits	7	3.5	2.5
		10	5.0	2.3
		15	5.7	2.2
GOLD	31 bits	7	1.9	1.6
		10	2.0	1.6

3.3.2 Training sequences

The product $\mathbf{M}^H\mathbf{M}$ in Eq. (12) is the correlation matrix of all sequences including both auto- and cross-correlation terms. Unfortunately, the inversion of the product $\mathbf{M}^H\mathbf{M}$ leads to the noise enhancement which limits the performance of a particular sequence set. The SNR degradation d_{ce} can be directly obtained from the diagonal elements of the matrix $(\mathbf{M}^H\mathbf{M})^{-1}$ and is given by [Stei94]

$$d_{ce} / dB = 10 \cdot \log_{10}\left[1 + \mathrm{tr}\left\{\left(\mathbf{M}^H\mathbf{M}\right)^{-1}\right\}\right]. \tag{13}$$

In the current GSM, the training sequences have ideal periodic auto-correlation functions over six or seven symbol shifts depending on the sequence which means that the noise enhancement is avoided in case of single signal channel estimation. For JCE, the noise enhancement cannot be totally avoided, as the cross-correlation properties are also counted. In the GSM training sequence set, the cross-correlation performance of the pair four and five from [ETSc] is very poor. A reason for this is that the sequences turn out to be reciprocal of each other.

Table 2 shows the performance of the GSM set according to the SNR degradation criteria of Eq. (13). In addition, the results of the length 20-bit sequences proposed for GSM in [Puk97] are given in the same table. This sequence type fits into the current GSM frame structure if the number of guard symbols is reduced from ten to six. For reference also the performance of the well-known length 31-bit Gold sequences is given. Gold sequences are known to have good correlation properties, but unfortunately they do not fit into the current GSM burst structure without modifications. From the table we find that for the set size seven, the best GSM pair is almost as good as the best pair of 20-bit sequences but when comparing the worst pairs 20-bit sequences outperform GSM sequences (worst (eighth) GSM sequence already left out). When the set size of 20-bit sequences is enlarged the performance of the worst pair of the 20-bit sequences is getting worse. Note that Gold sequences, although performing extremely well have the relative advantage of the longer sequence length as well as the larger basic set.

3.3.3 Identification of the dominant interferer

Identification of the dominant interfering signal on a burst-by-burst basis is mandatory in the DI cancellation especially with frequency hopping as the interference source will change from burst to burst. The identification can be accomplished as an integral part of the channel estimation process taking advantage of the different training sequences allocated to cochannels. An optimum DI identification method would estimate all the cochannels jointly using Eq. (12) and select the most powerful interfering signal. However, due to the fact that single valued solutions for Eq. (12) do not exist when the number of estimated channel parameters exceeds reference sequence length, this approach may not be feasible in GSM. For example, in case of four estimated channel taps, only four or five cochannels can be estimated simultaneously for reference sequence lengths of 16 and 20, respectively.

The most straightforward suboptimum method for DI identification is to measure the signal power after a filter bank matched to transmitted training sequences and select the one with the largest output power. However, due to the still rather strong cross-correlation values and relatively high power dynamics of the cochannel signals this approach may not provide a satisfactory performance. A more efficient method described in [Puk97] called pairwise channel estimation (PCE) method considers only two sequences at a time in computing Eq. (12) with the first sequence being always the desired signal training sequence, which is known a priori, while the second sequence is a candidate DI training sequence. All the interference training sequences are scanned and finally the best pair is selected either based on direct power estimation from the impulse responses or MSE criterion, i.e. the pair which minimises the residual signal

Figure 6. The probability of DI identification as a function of DIR i.e. I1/I2

$$\varepsilon = \left\| \mathbf{y} - \mathbf{M}\hat{\mathbf{h}} \right\|^2 \tag{14}$$

where $\mathbf{M}$ and $\hat{\mathbf{h}}$ include information of two sequences at a time. The advantage of the PCE method is that the cross-channel interference can be removed between two sequences.

Fig. 6 shows simulation results of the probability of DI identification as a function of dominant to rest of interference ratio (DIR) or I1/I2 ratio for the different algorithms as well as training sequence types. The training sequence set size is seven in all plots. Larger set sizes would slightly decrease the performance. Each signal is independently lognormally block fading and experiences fixed ISI in each block. Channel taps are (0.7479, 0.2441, 0.008) one symbol from each other. A lognormal power distribution has been used for interfering signals I1 and I2 and they are varying independently burst-by-burst basis modelling the frequency hopping (see Sec. 2).

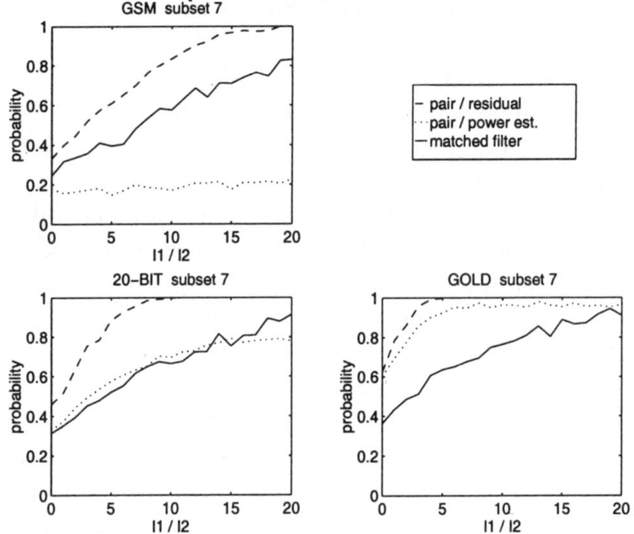

It turns out that the PCE method performs significantly better than the other methods. The reason for this may be that the signals excluded in the channel estimation act as a coloured noise which can be better combated by the MSE criterion using also the phase information of channel taps. With MSE criterion DI can be identified with 90-100% probability if DIR is greater than 5 dB. The performance somewhat varies with different sequence sets. The performance of the matched filter method is quite stable for all sequence types, unlike the performance of the PCE method depends very much on the used sequence set. Note that the curves present the average performance of all sequences in a subset, modelling the case of frequency hopping. It should be remembered that for different sequence pairs the performance may vary, which may be of interest, e.g., in non-frequency hopping cases.

3.4 Receiver complexity

3.4.1 Joint Detector

For joint detector, the main computational burden comes from the increase in the trellis size and transitions per state in the Viterbi algorithm. The number of trellis states is increasing exponentially by 2^{NL} and number of transitions per state is increasing as 2^{N}, where N is the number of cochannel signals in process and L is the channel memory length. To keep the complexity of the receiver tolerable, the advantage of suppression of DI alone is clear as N=2.

Compared to the conventional receiver, the trellis size of the Viterbi algorithm expands from 8 to 64 states or 16 states to 256 states in case of DI cancellation. The trellis size of 64 (4 channel taps per signal) is sufficient in urban areas and in rural, micro and indoor areas even a lower number of trellis states may suffice. In addition, the number of transitions per trellis state is increased from two to four compared to the current GSM receiver. The impact of this is that 1) number of compare-select operations per state is increased from one to three and 2) number of branch metric computations is doubled. Fortunately, the computation of a single branch metric value do not require extra processing power thanks to the block fading channel which allows us to precompute a look-up table of the possible channel output values (Eq. (4)) for a burst. Nevertheless, factor of eight to ten increase in complexity may be expected when updating the current GSM 16-state MLSE equaliser with 64-state JMLSE receiver implying that ASIC may be the most feasible approach for implementation.

In the literature a number of proposals have been made to reduce the complexity of the Viterbi equaliser. The most straightforward is to use the delayed decision feedback sequence estimation (DDFSE) truncating the channel impulse response and using only the truncated part to construct the

Viterbi trellis [Duel88, Eyu88]. The feedback part is used in the branch metrics computation. The complexity increase of this algorithm is only moderate with increasing number of channel taps. However, DFE structure is known to have difficulties in non-minimum phase channels, for which reason a prefilter turning the channel into minimum phase might be required [Mou94]. A method to reduce trellis states is to combine those states close to each other [Wal95]. Another method used in [Clar78] proposes to exclude the paths with low probability and keep only a fixed amount of paths for further processing.

3.4.2 Joint channel estimation

JCE increases the complexity of the receiver, but it can be reduced significantly when the product of the first three terms in Eq. (12) are kept in the receiver memory. In addition, the fact that the product consists of real valued terms can exploited when computing the complex algebra. Roughly, JCE doubles the computational burden compared to the single channel estimation. However, DI identification process increases the complexity by the factor number of training sequences minus one since the channel estimation has to be repeated for each desired signal and interference signal pair.

3.5 Other receiver issues

The problem of frequency offset between cochannel carriers may cause degradation in the receiver performance. The accuracy of BS frequency reference are 45 Hz for GSM and 90 Hz for DCS1800. In addition the Doppler shift might be in opposite directions for the interference and desired signal. This may result to maximum of 200-300 Hz frequency offset in the worst case which can be compensated by standard channel tracking algorithms, if necessary.

4. PERFORMANCE ANALYSIS

4.1 Simulation model

The DIR distribution has a major effect on the performance of the DI cancellation, therefore we have designed a novel link simulator in the performance analysis introduced in [Ran97b]. As shown in Fig. 7, the simulator includes a large number of interfering signals and each of them undergoes independently the multiplicative effects of multipath channel (fast fading) and lognormal fading. Both fading types are assumed to be independent from burst to burst modelling behaviour of ideal frequency hopping. The link simulator allows to evaluate and test following aspects important for joint detection:

- IC-gain relative to the conventional receiver
- DI identification algorithm
- required size of training sequence
- effect of the DIR distribution on IC-gain
- effect of frequency hopping

In case of hexagonal omnicell layout, the mean value of the lognormal distribution is defined by the average distance attenuation from interfering mobiles to the centre cell or vice versa. The standard deviation of the lognormal shadowing is obtained from the network simulator introduced in Sec. 2.2 with parameters introduced in Table 1 except no DTX is used. To justify our assumption of lognormal interference, Fig. 8 plots the interference distributions originating from a single cochannel cell for tiers 1, 2 and 3. It can be seen that the distributions look fairly Gaussian with some asymmetricity. The standard deviations of all interfering signals are ca. 11 dB and the mean value is proportional to the distance from the centre cell. Accordingly, these values are adopted in the link simulator. Note that a lower base station activity (DTX, load) will increase the standard deviation which can be controlled in the link simulator by randomly switching on/off interference bursts. The wideband channel model is Typical Urban and the GSM transmission parameters are used in all simulations [ETSc].

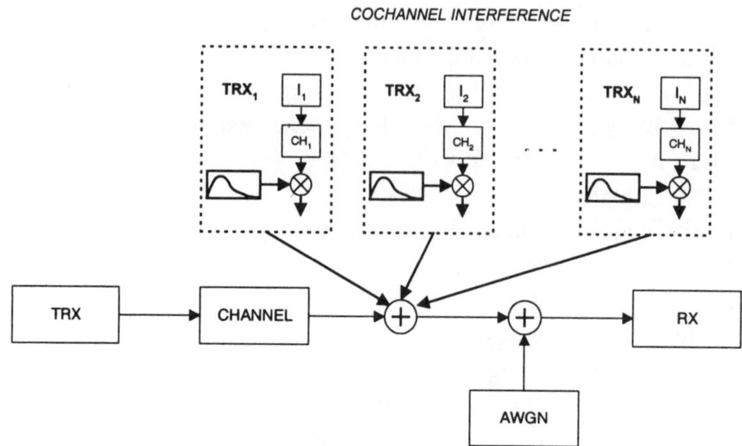

Figure 7. Lognormal fading simulation model

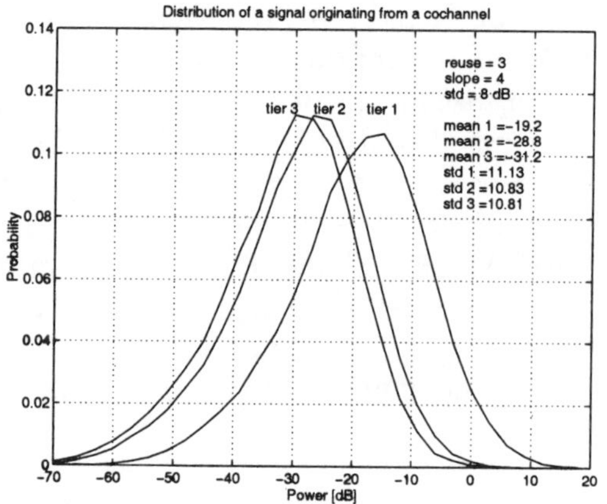

Figure 8. Interference distribution of a single cochannel interferer in tiers 1, 2, and 3.

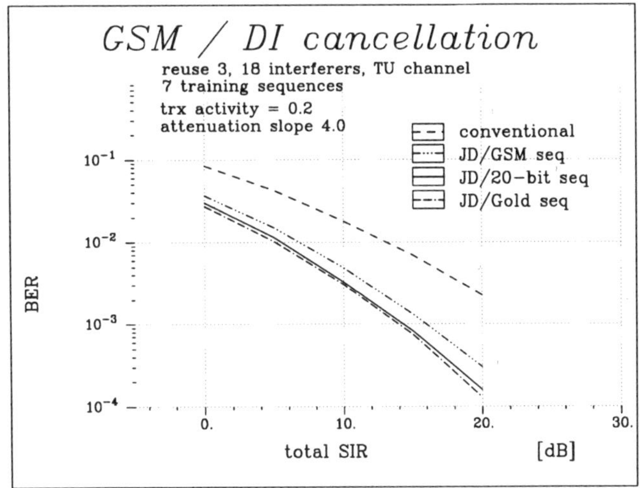

Figure 9. Performance of different training sequences

4.2 Results

In this section we compare the performance of different training sequence sets and the PCE DI identification algorithm. In addition, we investigate the effect of base station activity factor and cochannel asynchronism as well as the gain of suppressing two interferers. In all simulations uncoded BER is evaluated.

4.2.1 Training sequence performance

The performance of different training sequence sets is shown in Fig. 9. Seven different training sequences are allocated for 18 interferers such that the closest cochannel tier has different training sequence codes. As DI is changing randomly, the shown curve is an average performance of all sequences. We can see that the GSM sequences are 1.3 dB worse than 20-bit sequences, and furthermore, the Gold sequences do not seem to improve the performance very much. Results also indicate that training sequence set size of seven is sufficient in the omnicell case. In case of sectorised cells, even a lower number of training sequences might suffice.

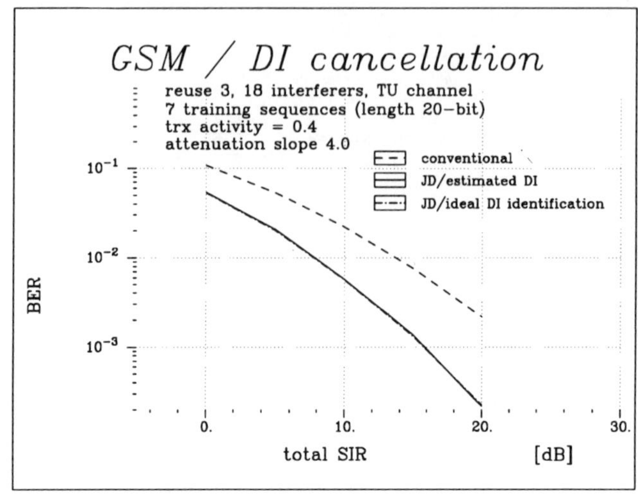

Figure 10. Performance of the DI identification algorithm

4.2.2 Performance of the DI identification algorithm

The performance of the PCE algorithm (see Sec. 3.3.3) is compared to the perfect DI identification in Fig. 10. We can see that there is negligible performance loss because of the estimation of DI. An explanation for this extremely good performance is that DI is very probably found when DIR > 5 dB and for lower values of DIR the IC-gain would be small anyway.

4.2.3 Effect of base station activity

Base station activity changes the standard deviation of lognormal shadowing which also influences to the DIR distribution. In Fig. 11 we plot BER curves for different base activity factors. We can see that the relative advantage of the JD/IC receiver increases with lower base station activity. We can see that the gain varies between 4 and 9 dB at BER 10e-2 depending on the base station activity. The upper value is reached at base station activity factor of 0.2 and the lower value with the factor of 1.0. A typical base station activity is 30-40 % (50% DTX included) which implies a gain of around 6 dB.

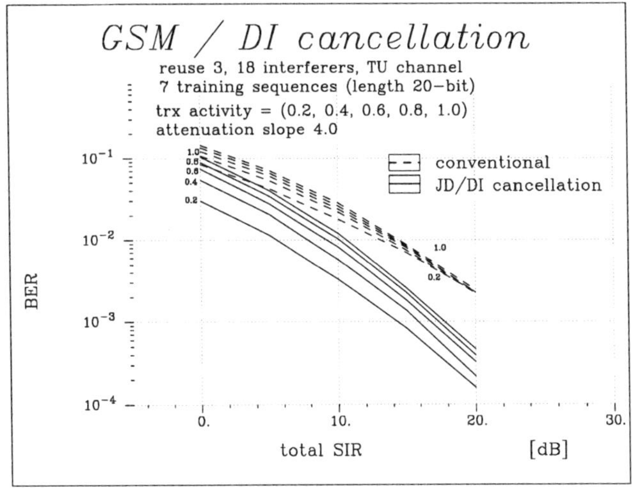

Figure 11. Effect of base station activity

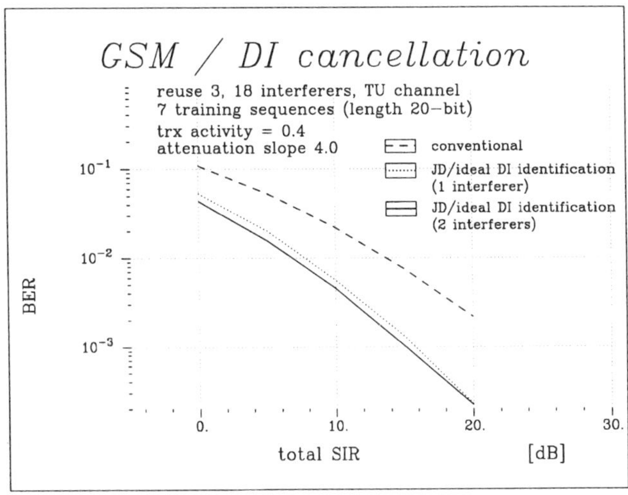

Figure 12. Performance of demodulation of two interferers with perfect DI identification

4.2.4 Suppression of two interferers

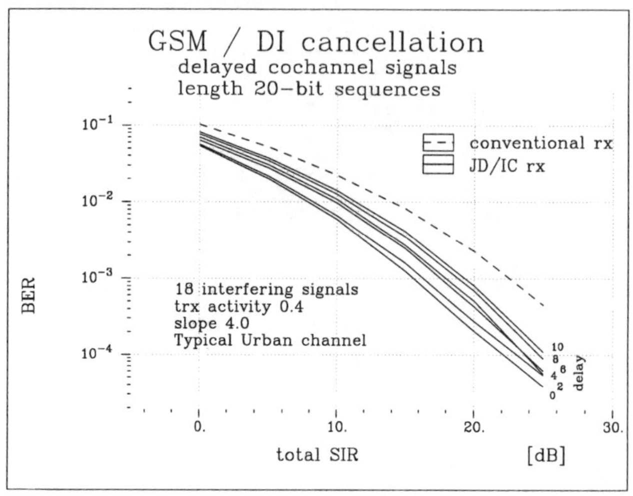

Figure 13 Effect of cochannel asynchronism. Time offset between desired and closest cochannel tier is varied from zero to ten symbols.

The performance of suppressing two interferers instead of one is plotted in Fig. 12 using ideal DI identification. For all cochannels four taps are used in the detector corresponding to 512 trellis states. We can see that the gain is only 1 dB indicating that the second largest interferer is often with much lower power than the largest one. Thus, the gain will be relatively small.

4.2.5 Effect of cochannel asynchronism

Although the base stations are completely synchronised, the propagation delay between cochannel signals will cause asynchronism between cochannel signals. This will degrade the receiver performance due to the lack of training sequence guard bits and tail symbols as well as worsened training sequence correlation properties. In Fig. 13 the relative time offset between desired and the first tier of interfering signals is varied from zero to ten symbols. The second and third tier of interfering signals have double as much time offset as they are even further off from the desired signal. In the channel estimator, time offset is assumed to be known which is taken account in the algorithm. In practice, the time offset should be estimated which may not critical as the time offset values do not change rapidly. From the results we find that the performance is gradually decreasing. Time offset of two symbols corresponds to 0.3 dB loss in performance as the offset of four symbols causes loss of 2 dB. Note, however, that in less severe multipath channels the degradation would be smaller.

5. SYSTEM REQUIREMENTS

5.1 Base station synchronisation

As indicated in Sec. 3.3, to support the proposed channel estimation method base station synchronisation is required enabling training sequences from different cochannels being received overlapping in time. The amount of tolerated asynchronism will depend on the training sequence design as well as maximum multipath delay. Synchronous systems also guarantee that the interference source is not different at the both ends of the burst which would make the interference suppression even more difficult.

There are several options how the base station synchronisation can be achieved. GPS (Global Positioning System) is a well known solution for outdoor cells and is also used by the IS-95 standard. The price of GPS receivers is nowadays reasonable but GPS has an disadvantage of not providing very good indoor coverage. GPS offers very accurate synchronisation, which on the other hand is not necessarily required by this application. Another method is to obtain base station synchronisation is to monitor the neighbour cell beacon signal, i.e. the BCCH carrier synchronisation sequence. If the cell sizes are small, the propagation delay causes only a small error in synchronisation accuracy and synchronisation can be obtained directly. For larger cell sizes and to obtain better synchronisation accuracy the propagation delay can be eliminated if

1. the distance between base stations is known or
2. base stations measure the time offsets between their own and the neighbour base station synchronisation sequences and report the results to a common node, e.g., to a Base Station Controller (BSC) or Mobile Switching Centre (MSC). The common node can compute the required time correction in the BSs' reference clocks.

5.2 Cell sizes and reuse factors

Although the base stations are synchronised, the cochannel signals experience different propagation delays which will cause asynchronism. To restrict the maximum propagation delay the reuse factors and cell sizes are limited. This is not a major problem since interference suppression is mostly needed in high capacity urban areas which inherently apply rather small cell sizes due to the requirement of high capacity and rather steep signal attenuation. Also usage of frequency hopping and DI cancellation themselves manifests a lower reuse in the network.

In Sec. 4.2.5 it was shown that the JD/IC receiver performance is gradually degrading when time offset increases from zero to ten symbols. In Fig. 14, the propagation delay between cochannel signals is investigated in a reuse three network with hexagonal cells. The example is described in the downlink direction and MS is located on the cell border being the most interesting location from the IC point of view. When the radius of hexagon is 1 (from the centre to a corner), the distance separation between interfering and desired signal paths is

$$I_{dist} - D_{dist} = \sqrt{21}/2 - \sqrt{3}/2 \approx 1.43.$$

The GSM symbol length is 3.69 us corresponding to 1.107 km propagation delay. In the above example 1 km cell radius means asynchronism of 1.3 symbol periods between the desired and interfering signals which causes negligible loss in the JD/IC-receiver performance (see Sec. 4.2.5) . If the cell radius was 5 km, the loss in interference cancellation gain would be 2 dB.

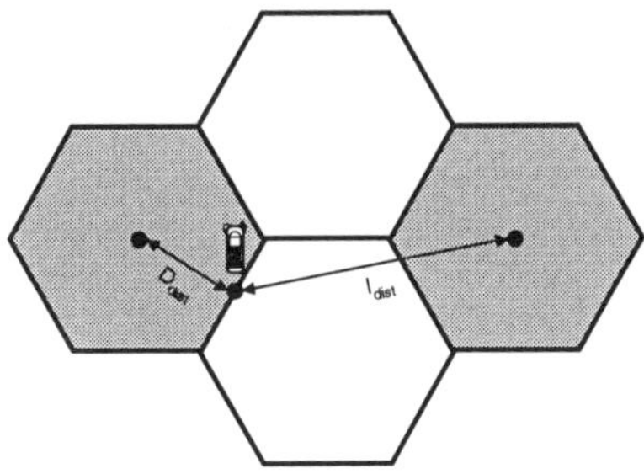

Figure 14. Worst case propagation delay between cochannels at the cell border (reuse 3 system).

5.3 Training sequences

Simulation results in Sec. 4.1.2. confirm that no more than seven distinct training sequences are required for omnidirectional cellular systems. In case of sectorised cells, even a smaller number of sequences might be enough. The average performance of the current GSM sequences is relatively good but still 1.3 dB improvement can be obtained by using the proposed 20-bit sequences. The advantage of the 20-bit sequences will probably be smaller when a smaller set of sequences suffice, e.g., in case of sectorised cells. The 20-bit sequences fit into current GSM frame structure, but usage of them requires a change in the standard.

The allocation of different training sequences can be done manually on the cell basis but it can be made automatically according to the measurements done in base stations. It is possible even consider automatic change of training sequences on the call basis or even during a call.

5.4 Control channels

If the performance of traffic channels can be improved with CCI, it is necessary to be able improve the performance of control channels accordingly. This is not a problem in those GSM control channels using the same training sequences as the traffic channels, e.g., SACCH, FACCH etc., since they can gain from JD/IC in a manner similar TCH channels. On the contrary, RACH and FCCH and SCH do not use the GSM training sequences and may require algorithm redesign in the receivers or in the worst case algorithm redesign of the standard. One way to avoid the problem is to guarantee lower interference level for BCCH and RACH timeslots by network planning which can be accomplished rather easily thanks to the BSs synchronisation.

6. APPLICATIONS

From the capacity point of view, due to the lack of antenna diversity reception in the downlink, it is clear that JD/IC technique should be used especially in the mobile receivers. When considering the application JD/IC technique in a current GSM network, it is evident that most of the mobiles will not be able to support JD/IC technique. Thus the reuse need to be adjusted according to performance of conventional receivers. Anyhow, there are at least five ways to gain from interference cancellation:

1. provide lower transmission powers for JD/IC mobiles and thereby generate less interference for the others.
2. enlarge the cell sizes of JD/IC mobiles
3. allocated dedicated carriers or timeslots for JD/IC mobiles and do separate frequency planning for them with a lower reuse
4. give enhanced quality or a higher bit rate for mobiles supporting JD/IC technique.

If new networks are designed supporting only JD/IC mobiles right from the beginning, an advantage of JD/IC technique is that the frequency planning will be less critical as JD/IC can reduce the interference problem. In microcells, JD/IC technique can be used to solve the street crossing interference problem [Ran97a] and potentially to alleviate the problem of street corner quality drop. In any case, the provision of BS synchronisation is much easier in small cells such as indoor and microcell systems. An interesting alternative is to comply SDMA systems with the JD/IC receiver technique.

7. CONCLUSIONS

In this Section an overview of CCI cancellation by joint detection in GSM systems has been presented. It has been shown that CCI cancellation is feasible in the GSM system to enhance the performance of the future GSM networks. The presented network simulations confirm that in the GSM network there is a high probability of a dominant interfering signal which gives significant computational advantage with only a minor loss in performance.

The receiver performance is analysed with a novel simulation system including 18 interferers each representing an interferer from a hexagonal omnicell layout. The results show that GSM training sequences perform satisfactorily although performance can be improved by 1.3 dB using an alternative training sequence set. The training sequence set size of seven sequences seem to be enough in the omnicell system, and the set size might be further decreased in case of sectorised cells. If a smaller set size is used, the penalty of using current GSM sequences becomes smaller. The effect of base station activity was investigated with respect to receiver performance. Depending on the BS activity factor the relative IC-gain varies between 4 and 9 dB at BER 10e-2. It was also demonstrated that JD of the two strongest interfering signals gives only 1 dB improvement compared to DI cancellation alone.

In addition to new receiver algorithms, the main requirement which the JD/IC technique poses to the system is the base station synchronisation enabling joint channel estimation. In addition, some limitations are set also for cell sizes and reuse factors due to the asynchronism caused by the propagation delay between cochannel signals. Moreover, either manual or automatic allocation of distinct training sequences for the nearest cochannels is necessary. As a conclusion, the GSM standard as such supports rather well the application of JD/IC technique, but some changes maybe favoured e.g. in the training sequence or control channel side.

REFERENCES

[Abe70] K. Abend and B.D. Fritchman, "Statistical Detection for Communication Channels with Intersymbol Interference", *Proc. of IEEE*, Vol. 58, No. 5, pp.779-785, May 1970.

[Ant97] C. Antón-Haro, J.A.R. Fonollosa and J.R. Fonollosa, "Blind Channel Estimation and Data Detection Using Hidden Markov Models", *IEEE Trans. on Signal Proc.,* Vol. 45, No. 1, January, 1997, pp. 241-246.

[Bea95] N. C. Beaulieu et. al, "Estimating the Distribution of a Sum of Independent Lognormal Random Variables", *IEEE Trans. on Comm.,* Vol. 43, No. 12, December, 1995, pp.2869-2873.

[Ber96] R. Berangi, P. Leung, and M. Faulkner, "Cochannel interference cancellation for mobile communications systems", *Proc. IEEE International Conference on Universal Personal Communications (ICUPC)*, Sep. 29th- Oct. 2nd , Boston, 1995, pp. 443-447.

[Bot95] G. Bottomley, K. Jamal, "Adaptive arrays and MLSE equalisation" *45th IEEE Vehicular Technology Conference* , Chigaco, 1995.

[Cha66] R.W. Chang and J.C. Hancock, “On Receiver Structures for Channels Having Memory”, *IEEE Trans. on Inf. Theory*, Vol. 12, No. 3, pp. 463-468, Oct. 1996.

[Che92] S. Chen and B. Mulgrew, "Overcoming co-channel interference using an adaptive radial basis function equaliser", *Signal Processing*, Vol. 28, 1992, Elsevier Science Publishers B.V., pp.91-107.

[Clar78] A.P. Clark, J.D. Harvey and J.P. Driscoll, “Near- maximum likelihood detection processes for distorted signals”, *The Radio and Electronic Engineer,* Vol. 48. No. 6. pp. 301-309, June 1978.

[Due88] Alexandra Duel-Hallen and Chris Hegaard, “Delayed Decision-Feedback Sequence Estimation”, *IEEE Transactions on Communications,* Vol. 37, No. 5, pp.428-436, May 1989.

[Edw96] N. Edwardsson, “Studies of Joint Detection-MLSE in the GSM System”, Master’s Thesis, Royal Institute of Technology, IR-SB-EX-9607, Stockholm, Jan. 1996.

[Esc97] Marko Escartin and Pekka A. Ranta, “Interference Rejection with a Small Antenna Array at the Mobile Scattering Environment”, *proceedings of First IEEE Signal Processing Workshop on Signal Processing Advances in Wireless Communications (SPAWC)*, La Bastille, Paris, April 16-18th.

[ETSa] ETSI STC-SMG11, “AMR - Working assumptions”, Draft 0.0.4., Meeting #2, Sophia Antipolis, April 21-25, 1997.

[ETSb] ETSI TC-SMG, “Overall description of the GPRS radio interface; Stage 2”, GSM Recommendations 03.64, v.5.0.0, July 1997.

[ETSc] ETSI TC-SMG, GSM Recommendations, Series 5, March 1997.

[Ett76] W. van Etten, "Maximum-Likelihood Receiver for Multiple Channel Transmission Systems", *IEEE Trans. on Comm.,* February 1976, pp. 276-283.

[Eyu86]. M. Vedat Eyuboglu amd Shadid U.H. Dureshi, “Reduced-State Sequence Estimation with Set partitioning and Decision Feedback”, *IEEE Transactions on Communications,* Vol. 36, No. 1, pp.13-20, Jan. 1988.

[Fal93] D. D. Falconer, M. Abdulrahman, W. K. Lo, B. R. Petersen and A. U. H. Sheikh, "Advances in Equalization and Diversity fpr Portable Wireless Systems", *Digital Signal Processing*, No. 3, pp. 148-162, 1993.

[For72] G. David Forney Jr., "Maximum-Likelihood Sequence Estimation of Digital Sequences in the Presence of Intersymbol Interference", *IEEE Trans. on Inf. Theory*, Vol. IT-18, No. 3, pp. 363-378, May 1972.

[Gir93] K. Giridhar et. al., "Joint Estimation Algorithms for Co-channel Signal Demodulation", *in Proc. IEEE Int. Conf. on Commun. (ICC)*, Geneva, 1993, pp. 1497-1501.

[How93] I. Howitt, "Recent Developments in Applying Neural Nets to Equalization and Interference Rejection", *Virginia Tech’s 3rd Symposium on Wireless Personal Communications*, Blacksburg, June 9-11, 1993.

[Karl96] J. Karlsson, J. Heinegård, "Interference rejection combining for GSM", *in proc. of 5th IEEE International Conference Universal Personal Communications*, Boston, MA, Sept. 29- Oct. 2, pp.433-437.

[Lau86] Piere A. Laurent, “Exact and Approximate Construction of Digital Phase Modulations by Superposition of Amplitude Modulated Pulses (AMP), *IEEE Transactions on Communications,* Vol. 34, No. 2, pp.150-160, Feb. 1986.

[Li95] Y. Li, B. Vucetic, and Y. Sato, "Optimum Soft-Output Detection for Channels with Intersymbol Interference", *IEEE Trans. on Inf. Theory*, Vol. IT-41, No. 3, pp. 704-713, May 1995.

[Mos96] Shimon Moshavi, "Multi-user Detection for DS-CDMA Communications", *IEEE Communications Magazine,* Vol. 34, No. 10, Oct. 1996, pp. 124-136.

[Mou92] Michel Mouly, Marie-B. Pautet, "The GSM System for Mobile Communications", 1992.

[Mou94] C. Mourot, J. De Vile, R. Hopper, P. Ranta, "The ATDMA equaliser design and implementation", *in proceedings of RACE Mobile Telecommunications Workshop*, Amsterdam, May 17-19,1994.

[Puk97] Markku Pukkila and Pekka A. Ranta, "Channel Estimator for Multiple Co-channel Demodulation in TDMA Mobile Systems", *2nd European Mobile Communications Conference (EMPCC'97)*, Bonn, Sept 30 -Oct 2, 1997.

[Ran95a] P.A. Ranta, A Hottinen, Z.C. Honkasalo, "Co-channel interference cancelling receiver for TDMA mobile systems", *in proceedings of IEEE International Conference on Communications (ICC'95)*, Seattle, 1995, pp.17-21.

[Ran95b] P.A. Ranta, Z.Honkasalo and J.Tapaninen, "TDMA Cellular Network Application of an Interference Cancellation Technique", *in proceedings of 1995 IEEE Vehicular Technology Conference (VTC'95)*, July 25-28 Chicago, Illinois, USA, pp.296-300.

[Ran96] P.A. Ranta, A. Lappeteläinen, Z.C. Honkasalo, "Interference cancellation by Joint Detection in Random Frequency Hopping TDMA networks", *in proceedings of IEEE International Conference on Universal Personal communications (ICUPC'96)*, Sept. 29-Oct. 2, 1996, Cambridge, MA.

[Ran97a] Pekka A. Ranta and Antti Lappeteläinen, "Application of dominant interference cancellation in street microcells", *IEEE International Conference on Communications (ICC'97)*, Montreal, 8-12th June, 1997, pp.17-21.

[Ran97b] Pekka A. Ranta, Markku Pukkila, "Recent results of Co-channel Interference Suppression by Joint Detection in GSM", *to be published in 6th International Conference on Advances in Communications and Control,* Corfu, Greece, 23-27 June 1997.

[Ran97c] Pekka A. Ranta and Marko Escartin, "Dominant interference cancellation by adaptive antennas in GSM", *in proceedings of IEEE International Conference on Personal, Indoor, Mobile Radio Communications,* Helsinki, Sept. 1-4, 1997.

[Rob95] P. Robertson, E. Villebrun, and P. Hoeher, "A Comparison of Optimal and Sub-Optimal MAP Decoding Algorithms Operating in the Log Domain", *in proceedings of Internatiol Conference on Communications (ICC'95)*, Seattle, WA, June 18-22, 1995, pp. 1009-1013.

[Ste94] B. Steiner and P. Jung, "Optimum and Suboptimum Channel Estimation for the Uplink CDMA Mobile Radio Systems with Joint Detection", *European Transactions on Telecommunications*, Vol. 5, No. 1, Jan.-Feb., 1994, pp. 39-50.

[Stu96] Gordon Stüber, "Principles of mobile communications", Kluwer Academic Publishers, p. 665, 1996.

[Ung74] Gottfried Ungerboeck, "Adaptive Maximum Likelihood Receiver for Carrier Modulated Data-Transmission Systems", IEEE Trans. on Commun., Vol. COM-22, No. 5, pp. 624-636, May 1974.

[Wal95] S. W. Wales, "Technique for cochannel interference suppression in TDMA mobile systems", *IEE Proc. Commun.*, Vol. 142, No.2, April, 1995, pp.106-114.

[Win84] J. Winters, “Optimum Combining in Digital Mobile Radio with Cochannel Interference”, *IEEE Journal on Selected Areas in Communications*, Vol. SAC-2, No. 4, pp. 528-539, July 1984.

[Yos94] H. Yoshino, K. Fukawa, H. Suzuki, "Interference Cancelling Equalizer (ICE) for Mobile Communications", Proc. IEEE Int. Conf. on Commun. (ICC), New Orleans, 1994, pp. 1427-1432.

Chapter 3

Spectral Capacity of Frequency Hopping GSM

K. Ivanov, C. Lüders, N. Metzner, U. Rehfueß
Siemens AG ÖN MN

Key words: Frequency Hopping, Frequency Diversity, Interference Diversity.

Abstract: In the highly competitive mobile radio market the system resource spectrum is often shared between at least two system operators while the capacity and quality requirements are dramatically increasing. Random Frequency Hopping (FH) in digital F/TDMA systems like GSM and its derivatives is a method of expanding the bandwidth used by a traffic channel, thereby creating frequency and interference diversity effects. In this article, the capacity gain from the diversity effects of random FH in combination with Power Control (PC) and Discontinuous Transmission (DTX) is analyzed by means of combined link and system level simulations of the GSM full rate speech channel embedded in different network simulation environments and different network configurations.

1. INTRODUCTION

The success of any cellular system is critically dependent on the allocation of sufficient radio spectrum and the system inherent spectrum efficiency. More spectrum or greater spectrum efficiency would generally imply a less complex, hence less expensive infrastructure and better operational efficiency. Since spectrum is limited, system engineers focus their efforts on the design of spectral and cost efficient systems by optimal utilization of system inherent options. GSM[1] as a digital F/TDMA system offers the radio link control options Frequency Hopping (FH), Power Control (PC) and Discontinuous Transmission (DTX). In practice, greater spectrum efficiency in cellular networks can be achieved through a tighter frequency reuse scheme and a higher base station density. Increasing the number of base sites per unit area reduces the cell radius and hence increases capacity per unit area. This approach is employed by the micro cell layer in hierarchical networks. Tighter frequency reuse patterns cause, however, a strong co-channel interference which limits the spectrum efficiency. In this context the feature FH improves signal robustness against interfering signals. Diversifying the interference, FH, in particular in combination with PC and DTX, will increase cell capacity. The objective of our investigations is to evaluate the capacity gain of both 1/3 and 3/9 reuse schemes with random FH compared with a 4/12 conventional network structure without FH.

2. BASICS OF FREQUENCY HOPPING

Frequency Hopping means changing the radio frequency (RF) used by a logical channel at a prescribed rate. If the frequency change takes place for every new burst, the FH is called slow frequency hopping whereas FH with RF changes faster than the modulation rate is termed fast frequency hopping. In this report only slow frequency hopping is considered.

Cyclic FH follows a pre-defined list of frequencies in a regular order. Random hopping, however, randomly chooses the frequency out of a predefined set. Both cyclic and random hopping can be implemented in baseband (BH) or synthesizer hopping (SH) technique. BH requires a dedicated TRX for each RF within the hopping sequence, whereas in SH the transceiver can be tuned to the desired RFs. Baseband FH imposes no restrictions on RF combining equipment, whereas SH requires wideband RF combiners.

The effects of FH on cellular networks are basically twofold. The mobile radio channel is characterized by multipath propagation resulting in frequency selective variations of the reception power. Experienced deep fades in the useful signal energy over consecutive bursts strongly disturb the signal's quality. By changing the carrier frequency - *frequency diversity* - the resulting fast fading pattern will change as well. Since only isolated

[1] Throughout this chapter the term GSM also applies for the GSM derivatives DCS1800 and PCS1900.

bursts will be affected, the remaining errors can be effectively corrected by a combination of forward error correction coding and interleaving like in GSM. This will improve the transmission link quality for slow moving mobiles in particular. Due to an improved frequency utilisation, cyclic hopping provides higher frequency diversity than random hopping. Assuming a cyclic hopping mode over 8 frequencies, all 8 bursts within the interleaving length of one speech frame are transmitted at a different frequency, whereas in a random hopping mode only 5 frequencies are used on average.

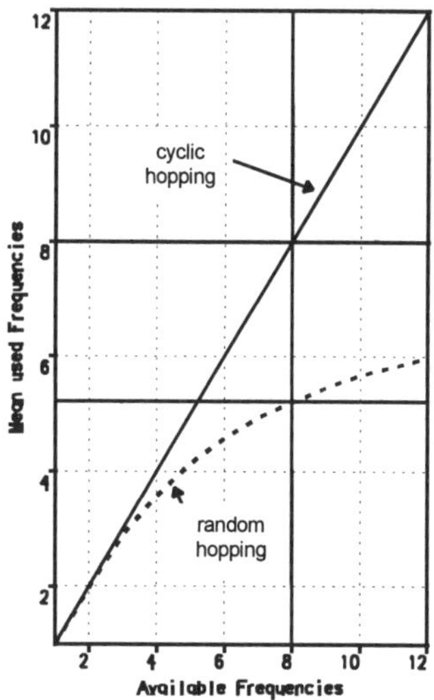

Figure 1. Frequency utilisation in cyclic and random hopping mode

In nonhopping or cyclic hopping systems, radio links between mobiles and base stations in cellular networks may suffer on several sequential bursts from co-channel interference generated by strong transmitters in the same frequency and time domain. When random FH is used, the co-channel interference experienced on a particular link is generated by different sets of transmitters and can therefore vary from burst to burst - *interference diversity*. A cyclic hopping system needs to be designed according to worst case interference conditions whereas a random hopping one with an average interference value. With DTX the average interference for each user will be reduced by the DTX factor. In a fixed or cyclic hopping system instead, the interference has an on/off nature with the maximum interference being the same as without DTX [2].

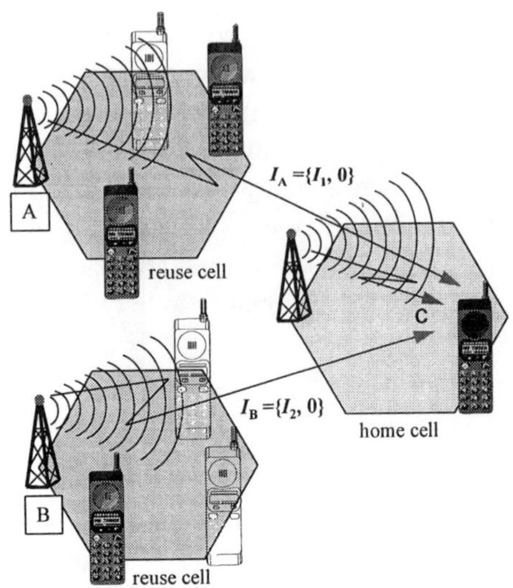

Figure 2. Example interference situation

To illustrate the effects of interference diversity, let us have a look at a simple example network given in *Figure 2*:

A mobile in its home cell receives a wanted signal C from the serving base station and interfering signals I_1 and I_2 from base stations A and B, respectively. Assuming hopping over 3 frequencies and 50% load, consider two active mobiles in cell A and one active mobile in cell B. PC and DTX are switched off. Cyclic hopping will result in 1 out of 4 different interference scenarios: $I = 0$, $I = I_1$, $I = I_2$ or $I = I_1 + I_2$ as shown in Figure 3. In any case, the interference remains constant throughout the whole duration of the call. With $n = 2$ interfering cells and $f = 3$ random hopping frequencies there are $f^n = 3^2 = 9$ different interference conditions. In this case the mobile in the home cell will face a different interference situation at each burst thus averaging the interference to $I_{mean} = (2 \cdot 0 + 4 \cdot I_1 + 1 \cdot I_2 + 2 \cdot (I_1 + I_2)) / 9$. Assuming $I_1 = 4 \cdot I_2$, the mean interference results in $I_{mean} = 3 \cdot I_2$.

Furthermore, interference diversity produces a second effect similar to that of frequency diversity. Since collisions of the carrier signal with strong interferers do not occur on consecutive bursts, errors from occasionally corrupted bursts can be effectively corrected by the forward correction scheme. This results in an improved transmission link quality independent of the mobile speed. Obviously the collision probability with strong

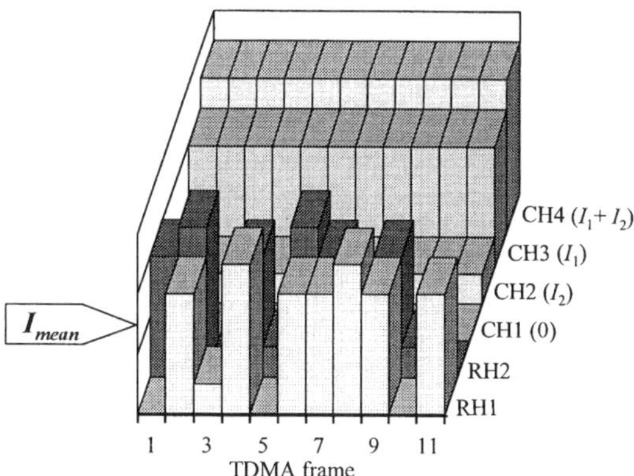

Figure 3. Interference situations for random (RH1..2) and cyclic (CH1..4) hopping

interferers will decrease as the number of hopping frequencies increases and the system load decreases. On system level this implies that the maximum gain from interference diversity will be expected in networks with tight frequency reuse and fractional load.

3. SYSTEM QUALITY IN FH-GSM

GSM employs slow FH, i.e. the carrier frequency of a logical channel is changed for every new burst (every 4.615 ms). FH can be performed over a maximum of 64 frequencies. Random hopping utilizes pseudo-random sequences with a length of 84864 elements [3]. The random hopping sequences in reuse cells are given different shifts producing uncorrelated hopping sequences, and thus the collisions between co-channel users will be random. Each hopping sequence can be used with a frequency offset resulting in as many sequences as the number of frequencies used in the hopping sequence. These sequences are orthogonal and can therefore be used within a cell to avoid intracell interference.

The channel coding schemes specified for GSM fully rely on interleaving distributing effectively the transmission of a speech frame in the time domain, i.e. over 8 consecutive TDMA-frames which is a time period of 32.88 ms.

In fully interference limited systems, i.e. the network is dimensioned to accommodate maximum capacity, GSM specifies a reference interference performance in terms of (*BER*s, *FER*) which is to be met for a mean *C/I* ratio of 9 dB [5]. This is termed soft blocking level.

For fixed and cyclic hopping systems the mapping between the mean *C/I* and the error rates (*BER*s, *FER*) on the radio link is unique, and it is independent of the system load. If the impairment of co-channel interference is studied, the required communication quality in such systems is reliably specified by a certain mean *C/I* value for a given outage probability (e.g. 9 dB for 10% outage in GSM) of area coverage [6]. However, in a system with f randomly hopping frequencies and n co-channel cells an MS at a particular location will perceive f^n different *C/I* values forming a stochastic process with certain characteristics. Due to the interference diversity effect a certain mean *C/I* ratio can result in quite different error rates depending on the characteristics of the stochastic process [7]. Thus the mean *C/I* ratio as a quality measure for random hopping systems is not adequate, since not only the average interference value determines the quality, but also the distribution of interference on different bursts.

Thus in a random hopping system, the error rates (*BER*, *FER*) after channel decoding shall be used instead. As a result, the quality on the speech channel is termed acceptable if in at least e.g. 90% of the service area the performance of the GSM full rate speech traffic channel (TCH/FS) fulfills the quality requirement in terms of e.g. $FER \leq 2\%$.

The focus of our evaluation is on the resulting effect of both frequency and interference diversity on the spectral capacity of a spectrum limited system in a dense urban scenario with slowly moving subscribers.

4. SPECTRAL CAPACITY OF GSM

For speech services the capacity χ required in a certain service area A of a mobile radio network is proportional to the traffic λ that can be served at a given quality of service (QoS) within this service area.

$$\chi = \frac{\lambda}{\mathrm{A}} \tag{1}$$

The following criteria are usually used to define the QoS:

Call setup phase: The blocking due to "*no resource available*" (hard blocking) shall be below a given threshold (e.g. less than 2%; usually the blocking is calculated by the Erlang formula).

Connected mode: In connected mode the service quality is usually characterized in terms of *C/I* ratio, *BER*, *FER* thresholds (soft blocking)

which have to be met in a certain portion (e.g. 90%) of the service area. In random hopping systems as explained above a *FER* threshold of e.g. 2% shall be met within e.g. 90% of the service area.
Expanding equation (1) yields:

$$\chi = \frac{\lambda}{N_{CH}} \cdot \frac{N_{CH}}{N_{RF}} \cdot \frac{N_{RF}}{N_{site}} \cdot \frac{N_{site}}{A} \tag{2}$$

with

N_{CH} number of channels,
N_{RF} number of carriers (RF channels),
N_{site} number of sites.

In cellular mobile radio systems the factor N_{RF} / N_{site} - number of carriers per site - is determined by the frequency reuse scheme which can be sustained at a given QoS. Using the definition of *cluster size* for homogeneous networks as the number sites (e.g. one BTS site may serve one omni or 3 sector cells covering approximately the same area) where each radio frequency carrier is used exactly once, the following equation can be derived:

$$\frac{N_{RF}}{N_{site}} = \frac{B \cdot \frac{N_{RF}}{B}}{K} \tag{3}$$

with

B [MHz] system bandwidth,
K cluster size.

Inserting (3) in (2) gives:

$$\chi = \frac{\lambda}{N_{CH}} \cdot \frac{N_{CH}}{N_{RF}} \cdot \frac{N_{RF}}{B} \cdot B \cdot \frac{1}{K} \cdot \frac{N_{site}}{A} \tag{4}$$

The factors constituting Eq. (4) are:

- *traffic/channel* (λ / N_{CH}): system load; this factor takes into account the fact that a channel usually cannot be fully loaded due to either hard or soft blocking.

- *channels/carrier* (N_{CH}/N_{RF}): system dependent parameter (GSM family: 8 TCH (traffic channels) for fullrate channels, 16 TCH for halfrate channels, signaling neglected);
- *carriers/bandwidth* (N_{RF}/B): system dependent parameter (GSM family: 5 carriers per 1 MHz);
- *bandwidth* (B): system bandwidth allocated to a network operator;
- *cluster size* (K): characterizing the frequency reuse in the deployment area, which depends on propagation conditions, required QoS and the network cell layout;
- *sites/area* (N_{site}/A): describes the base station density in the deployment area.

Eq. (4) shows the basic options for potential increase of the system capacity. The allocated spectrum depends upon regulatory issues and license costs while the site density is strongly related to the site acquisition costs. The other parameters involve the pure technical efficiency of the mobile radio system under consideration.

The *spectral capacity* η of a cellular network is obtained by relating its capacity χ to the most severe network investment costs spent for the licensed spectrum and building up of the network structure, i.e.:

$$\eta = \frac{\chi}{B \cdot \frac{N_{site}}{A}} \tag{5}$$

Replacing χ from Eq. (4) gives:

$$\eta = \frac{\lambda}{N_{CH}} \cdot \frac{N_{CH}}{N_{RF}} \cdot \frac{N_{RF}}{B} \cdot \frac{1}{K} \tag{6}$$

It should be noted that the definition of spectral capacity for speech services implicitly includes the speech coding efficiency, e.g. the spectral capacity for GSM half rate speech channels is twice as high as the one for full rate traffic channels assuming a comparable speech quality of both codecs.

Unfortunately the calculation of the spectral capacity based on Eq. (6) is not straightforward. The factors *channels/carrier* and *carriers/bandwidth* are system constants, whereas the *traffic/channel (system load)* and the *cluster size* have to be derived from system simulations (or combined system and link level simulations) and teletraffic theory (Erlang formula).

In a system without random FH there is no interference averaging and therefore, the worst case assumption has to be made concerning the impact

of interference, i.e. the *minimum cluster size* satisfying the QoS criterion for 100% system load in connected mode has to be found. The *maximum system load* is then calculated using the Erlang formula for the number of channels per cell resulting from the bandwidth licensed to the operator and the employed frequency reuse scheme (*minimum cluster size*). Such systems are operated at the *hard blocking limit.*
However, applying random FH in a cellular network provides interference averaging, allowing to design the system based on an average interference level instead of a worst case assumption of 100% system load. For a smaller cluster size, this implies a maximum system load limited by interference - *soft blocking limit* - which is usually lower than the load corresponding to the hard blocking limit mentioned above.
For a given cluster size, the maximum system load due to soft blocking at which the QoS (e.g. 2% FER in less than 90% of the service area) can still be met, is evaluated by means of system simulations including a statistical radio link model described in Section 5.
Keeping in mind the motivation for spectral capacity as a measure for the ratio between operator's revenue and investment costs, the decision, whether a soft blocking or a hard blocking configuration is more efficient, is based on the evaluation of the respective spectral capacity.
The spectral capacity evaluation process comprises several steps.

Step 1: Calculate the number of carriers per cell (N_{RFcell})

- hard blocking scenario

Usually the same reuse pattern is assumed for both the BCCH and the non BCCH carriers resulting in:

$$N_{RFcell} = \frac{N_{RF}}{B} \cdot B \cdot \frac{1}{K_{HB} \cdot N_{\sec}} \tag{7a}$$

with

K_{HB} hard blocking cluster size,
N_{sec} number of sectors (cells) per site.

For classical BTSs without wideband transceivers, the number of radio carriers per cell is equal to the number of transceivers per cell.

- soft blocking scenario

Introducing random frequency hopping, a tighter frequency reuse scheme (soft blocking cluster size) may be applied for the non BCCH carriers.

However the BCCH carrier has to be allocated using the hard blocking cluster size. Combining both reuse schemes yields:

$$N_{RFcell} = \frac{N_{RF}}{B} \cdot B \cdot \frac{1}{K_{SB} \cdot N_{\sec}} - \frac{K_{HB}}{K_{SB}} + 1 \tag{7b}$$

with

K_{SB} soft blocking cluster size.

Since for the soft blocking scenario the traffic per radio carrier is usually substantially lower than given by the Erlang formula, it is economically preferable to install less transceivers per cell than radio carriers available. This method requires synthesizer hopping.

Step 2: Calculate the number of channels per cell (N_{CHcell}):

$$N_{CHcell} = \frac{N_{CH}}{N_{RF}} \cdot N_{RFcell} \tag{8}$$

with N_{RFcell} either from Eq. (7a) or (7b) whichever is appropriate.

Step 3: Calculate the number of traffic channels per cell ($N_{TCHcell}$)

$$N_{TCHcell} = N_{CHcell} - N_{sig} \tag{9}$$

The number of required signaling channels (N_{sig}) is calculated by taking into account the signaling channel structure of the system and a traffic model (ratio between traffic and signaling load, e.g. 25 mErl / 5 mErl).

Step 4: Calculate the traffic per cell (λ_{cell}):

- hard blocking scenario

$$\lambda_{cell} = \mathrm{ErlangB}(blocking, N_{TCHcell}) \tag{10a}$$

with ErlangB denoting the Erlang-B formula applied on $N_{TCHcell}$ channels at a given *blocking*, e.g. 2%.

- soft blocking scenario

$$\lambda_{cell} = L_{SB} \cdot \tfrac{N_{CH}}{N_{RF}} \cdot \left(N_{RFcell} - 1\right) + \left(\tfrac{N_{CH}}{N_{RF}} - N_{sig}\right) \tag{10b}$$

with L_{SB} soft blocking system load (λ / N_{CH}) obtained from simulations. Note that the number of carriers per cell N_{RFcell} is calculated using Eq. (7b).
In Eq. (10b) the first term represents the traffic carried by the non-BCCH carriers in the tight reuse scheme (K_{SB}) while the second one is the contribution provided by the BCCH carrier allocated in (K_{HB}). The number of signaling channels N_{sig} required is estimated similarly to the hard blocking scenario by taking into account both terms in Eq. (10b).

Step 5: Calculate the traffic per site (λ_{site}):

$$\lambda_{site} = \lambda_{cell} \cdot N_{\sec} \tag{11}$$

Step 6: Calculate the spectral capacity (η):

$$\eta = \frac{\lambda_{site}}{B} \,. \tag{12}$$

5. INTEGRATED SYSTEM SIMULATION MODEL

The aim of the simulations is to evaluate the maximum system load L_{SB} fulfilling the QoS in different network configurations using random FH with different number of hopping frequencies.
Collecting statistics of the error rates required for channel quality estimation in a random hopping system requires a simulation environment including both system level and link level models. This could be achieved by keeping each MS in a system level simulation alive for some time, let it perform a number of hops generating C/I ratios which are fed into a link level simulation model. The outputs of the link level simulation are error rate estimates for this particular mobile. However, to limit the complex link level simulations for several thousands of mobiles a Statistical Radio Link Model (SRLM) has been designed. The SRLM has been embedded into a system level simulation environment referred to as Network Model (NWM) providing an Integrated System Simulation Model (ISSM). As a system option, FH is implemented within the NWM using the Hopping Sequence Generator as specified in [3].

The design of the Statistical Radio Link Model is based on a statistical analysis of the results produced using a common link level simulation model. In this model the GSM FEC (foreward error correction scheme) has been implemented according to [4]. At the transmitting front end, a full rate speech encoder delivers a data block of 260 bits corresponding to a 20 ms speech frame. The bits produced by the speech encoder have different importance and are classified into class Ia, class Ib and class II bits. The channel coding introduces redundancy into the data flow for error correction purposes. For the channel type TCH/FS, firstly a block code is applied to the class Ia bits (for frame erasure detection), and secondly all class Ia and class Ib bits are encoded by a convolutional code ($r = 1/2$, $K = 5$). The class II bits remain unprotected. The reordering and interleaving process mixes the encoded data blocks (frames) of 456 bits and groups the bits into 8 sub-blocks (57 bits = half burst). The 8 sub-blocks of one frame are transmitted on 8 successive bursts interleaved with the last 4 sub-blocks from the previous and the first 4 sub-blocks from the next frame, respectively (i.e. interleaving depth equals 8). Erroneously received bits within a particular frame, due to multipath propagation, are uniquely associated with these 8 bursts. The best error correction performance of the convolutional code is achieved when bit errors are randomly positioned. This is the reason why reordering and interleaving have been introduced in the GSM signal transmission flow. This implies that reordering/interleaving can only improve the FEC performance if the 8 successive bursts carrying the data of one frame are exposed to uncorrelated fading.

At the receiving front end, the modulated burst signal enters a Viterbi equalizer, see *Figure 4*. The demodulated bits with associated soft decision values are input to the deinterleaver, where the format of the encoded frame is restored from the 8 sub-blocks received on 8 successive bursts. The encoded frame is then submitted to the channel decoder. If a frame is declared "good", the decoded bits are forwarded to the speech decoder. Both class Ib and class II RBE are evaluated for good frames only.

The task of the SRLM is to map the C/I_{burst} onto BER, FER and class IbRBER values at the receive front end. Reflecting this task *Figure 4* shows the simplified block structure of the GSM TCH/FS receiver model and the corresponding blocks of the SRLM.

The function performed by each block of the SRLM can be described by statistical input/output relations of the corresponding blocks in a common link level simulator. To study these relations, statistics have been collected from link level simulations assuming a fully interference limited system scenario for the GSM TCH/FS in a TU3 propagation environment with ideal frequency hopping. In the link level simulations a frame was declared "bad" if at least 1 out of the 50 class Ia bits has been erroneously received.

The first block (C/I_{burst} onto BE_{burst} mapping) provides a statistical relationship between the C/I_{burst} and the resulting number of BE_{burst} (Bit Errors on burst). The second block (deinterleaving) of the SRLM determines the cumulative number of bit errors per frame BE_{frame}. The output of the de-interleaving block (BE_{frame}) is piped to the third block (FE decision and class IbRBE estimation) of the SRLM. The decision as to whether a received frame is "bad", i.e. a Frame Erasure (FE) occurs, or "good" is based on a threshold comparison of the number BE_{frame} with a statistical quantity referred to as Frame Erasure Decision Threshold (FEDT). The estimation of the number of class IbRBE is accomplished by

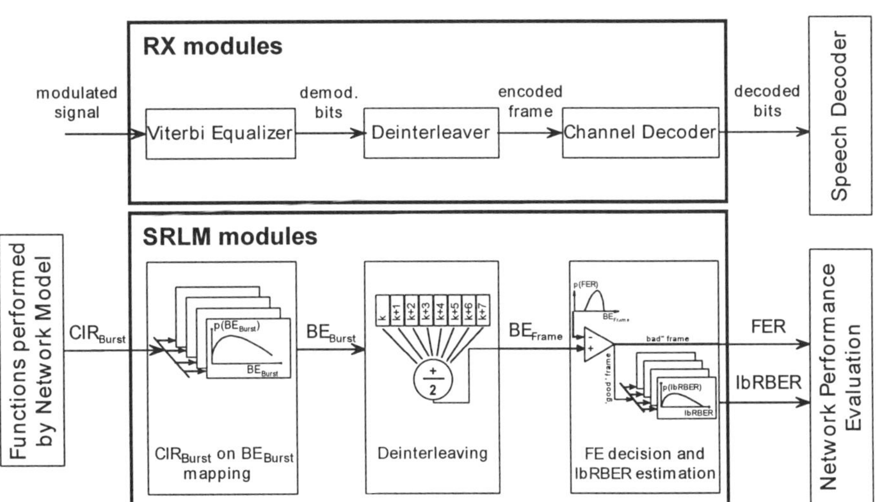

Figure 4. Model structure of the GSM transmission path at the receive end and the corresponding blocks of the SRLM

mapping the number BE_{frame} onto a *class IbRBE* using the statistical relationship between these quantities derived from the link level simulation results.

The NWM establishes the network configuration, positions the mobiles (MSs) at random locations in the cell area of the home and reuse cells, and calculates the received local mean power of the carrier and interfering signals by considering a certain path loss and slow fading signal variations due to shadowing. Furthermore the NWM affiliates each MS to a proper serving BS according to a particular cell selection criterion (e.g. best received level) and afterwards accomplishes the PC process. In addition, for each radio link a fast fading value is read in from a stored data file for the multipath propagation profile (e.g. TU3, TU50) under study. In the FH case it is assumed that the hop size is greater than the coherence bandwidth of the propagation channel. The NWM calculates the C/I_{burst} measured at the home MS or BS on each burst.

The C/I_{burst} values are used by the SRLM to assess the radio link quality in terms of *BER*, *FER* and *class-IbRBER* as observed at each particular MS location.

Finally, the results obtained for a sufficient number of MS locations are used by the NWM to generate the CDFs of *BER*, *FER* and *class IbRBER* describing the network performance. These CDFs are used to estimate the maximum system load L_{SB} which assures that at least 90% of the calls (MS locations) will meet the QoS criterion.

6. SIMULATION SETUP

The network consists of one central cluster of cells surrounded by three tiers of interfering clusters. Three-sector base sites with clover-leaf cell layouts are assumed with varying reuse patterns (1/3, 3/9 and 4/12). Based on the Monte Carlo method, the Integrated System Simulation Model is run a sufficient number of simulation cycles (e.g. 10000). At the beginning of each simulation cycle new MS locations in the home and reuse cells are generated uniformly distributed over the respective cell areas. The MSs remain stationary during the simulation cycle, i.e. the path loss and lognormal fading will not change during a simulation cycle. The slow fading values (normally distributed in dB) are assumed independent for the wanted signal and all interfering signals. Thus the total interference power is the sum of the power of all active interferers. The serving BS for each MS is selected using a cell selection criterion based on minimum pathloss.

Table 1. Simulation Parameters

Parameter	Value
Network Structure	19 clusters, 3-sector base sites with clover-leaf cell layout, reuse pattern 4/12, 3/9 and 1/3
Antenna	66° 3dB-beamwidth, 120° 10dB-beamwidth
Cell Radius	1000 m
Cell Selection / Handover	based on best received level, handover offset 5 dB
Probe Call Duration	8 s at each single MS location (400 frames)
Power Control	signal level based, enabled for reuse pattern 3/9 and 1/3, disabled for reuse pattern 4/12
Pathloss model	L [dB] = 18.8 + 38 · log(d [m]) (GSM 03.30 [5] for small cells)
Std. Dev. of Log-Normal fading	7 dB (according to GSM 03.30 [5])
Multipath Propagation Profile	TU3 (uncorrelated fading on different frequencies)
Frequency Hopping Sequences	uncorrelated between cells, orthogonal within cells
DTX factor	0.6

The effect of handover offset is taken into account by randomly selecting the serving cell out of a set of candidate cells. The possible set of base stations to be affiliated to is defined as the set of all base stations that has a path loss not greater than a handover offset (e.g. 5 dB) higher than the pathloss to the base station with the lowest path loss. GSM compliant slow PC is implemented based on received signal level [8].

Fast fading is time varying according to the terminal speed defined in the multipath propagation model representing a terminal moving in a small area around its fixed position. During each simulation cycle a sufficient number of speech frames (400 are found to provide a good convergence) are transmitted on the radio link. The downlink quality is used as a performance measure since it generally limits the system due to the lack of antenna diversity at the MS receiver.

The important parameters used in the simulations are summarized in *Table 1*.

The simulation model assumes a synchronized and fully co-channel interference limited system. Adjacent channel interference is not considered. Since the main goal of this report is to study the implications of FH on the design of spectrum and cost efficient systems with limited spectrum allocation, a typical deployment scenario is assumed to involve a dense urban environment with stationary or very slowly moving MSs.

7. SIMULATION RESULTS

To study the resulting diversity effects of random FH as a function of the system load L_{SB} and the number of hopping frequencies, simulations with various reuse schemes have been performed. First we studied the influence of the system load. *Figure 5* shows the cumulative density functions (CDFs) of *FER* with 8 hopping frequencies per cell in a 1/3 clover-leaf reuse pattern. For a low system load (25%) the coverage probability at 2% FER improves significantly, benefiting from the channel coding gain due to interference diversity. The bold vertical line at 2% *FER* indicates the threshold of the QoS criterion. The target GSM performance of 90% area coverage (horizontal line) is exceeded.

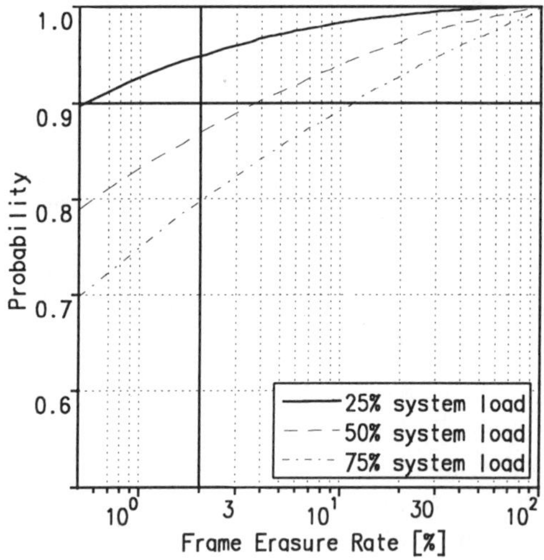

Figure 5. CDF of FER of a random FH system over 8 frequencies with varying system load in a 1/3 clover-leaf reuse scheme

We then studied the effect of the number of hopping frequencies on the system performance. The results in *Figure 6* obtained for a system load of 25% indicate that even when employing random hopping over 2 frequencies the GSM QoS criterion ($FER \leq 2\%$) can be met. As expected, the gain is increased further by adding more frequencies to the hopping sequence.

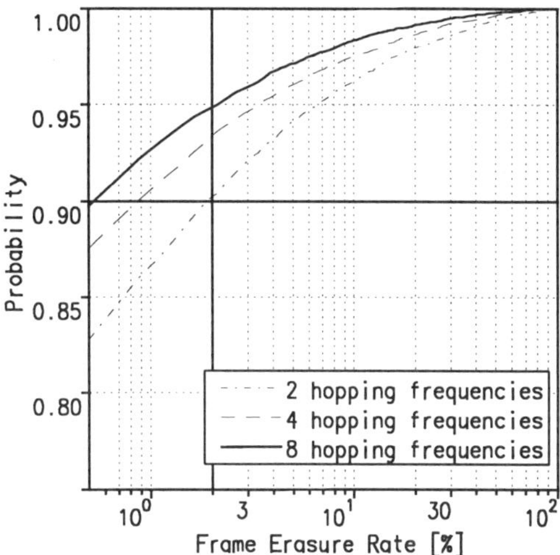

Figure 6. CDF of FER of a fractionally loaded (25%) random FH system with varying number of hopping frequencies in a 1/3 clover-leaf reuse scheme

However, simulations carried out with more than 16 frequencies have shown that (keeping the system load constant), the resulting quality improvement shows a saturation effect.

As a reference, *Figure 7* shows the coverage probability in a conventional

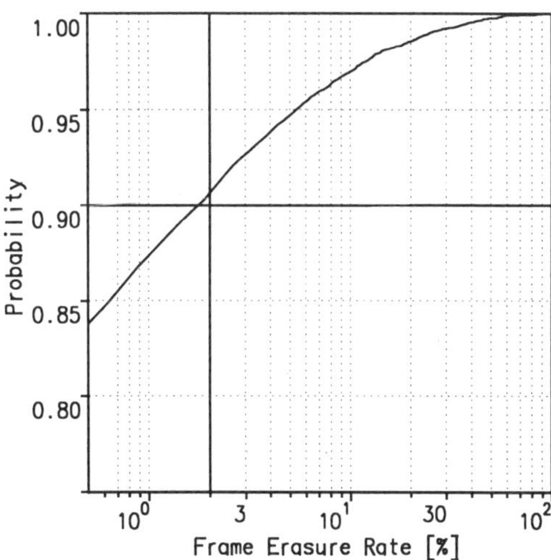

Figure 7. CDF of FER of a fully loaded non-hopping 4/12 clover-leaf reuse scheme without PC

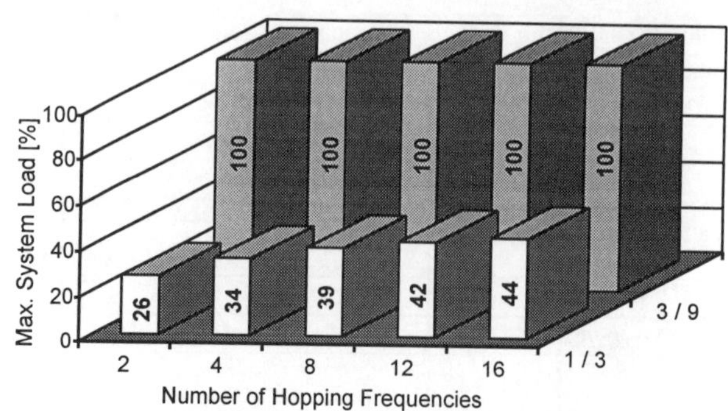

Figure 8. Maximum system load due to soft blocking as a function of the number of hopping frequencies for reuse schemes 1/3 and 3/9 with PC but without DTX

non-hopping 4/12 reuse scheme without PC as utilized for the BCCH carriers.

To determine the maximum system load due to soft blocking L_{SB} for various reuse schemes, we varied the system load for fixed number of frequencies. As an example, for a 1/3 clover-leaf reuse pattern with 8 hopping frequencies in each cell the maximum system load was found L_{SB} = 39%. The obtained performance is depicted in *Figure 8* and will of course improve further with DTX.

Figure 8 shows the maximum system load L_{SB} for a clover-leaf network configuration with a frequency allocation scheme of 1/3 and 3/9, respectively. Obviously, a 3/9 reuse scheme with PC is not interference limited when FH is applied regardless of the number of hopping frequencies employed.

Note, that DTX has not been taken into account in the maximum system load depicted in *Figure 8*. The effect of DTX is twofold. In a soft blocked reuse scheme like 1/3 the interference reduction from DTX can directly be converted into additional capacity due to an increased soft blocking limit. Employing DTX in a hard blocked reuse scheme like 3/9 will greatly improve system quality rather than increasing capacity.

As generally recommended a 4/12 reuse pattern is assumed for the BCCH frequencies. The values of the maximum system load L_{SB}, taken from the simulations correspond to the maximum number of frequencies available for random hopping in each cell of the respective reuse scheme. Applying DTX in a random hopping system, the actual maximum system load due to soft blocking is given by $\min(1, L_{SB}/\beta)$, with e.g. $\beta = 0.6$ as DTX-factor. The following example illustrates the calculation procedure as described in Section REF4 for the spectral capacity figures shown in *Figure 9*.

Table 2. Example of calculating the spectral capacity of hard and soft blocked reuse schemes, respectively

Scenario	4/12 at hard blocking	1/3 at soft blocking	Equation used
N_{Rfcell}	2	5	(7a), (7b)
N_{CHcell}	16	40	(8)
$N_{TCHcell}$	14	37	(9)
λ_{cell} [Erl]	ErlangB(2%, 14) = 8.2	0.57 · 8 · (5 - 1) + (8 - 3) = 23.2	(10a), (10b)
λ_{site} [Erl]	24.6	69.6	(11)
η [Erl/(site·MHz)]	5.1	14.5	(12)

Assume e.g. B = 4.8 MHz bandwidth allocated to a GSM network operator (N_{RF}/B = 5). Using only full rate traffic channels (N_{CH}/N_{RF} = 8) the results of the following scenarios are presented in *Table 2*:

- cluster 4 with 3 sector sites (4/12) at the hard blocking limit: (K_{HB} = 4, N_{sec} = 3)
- cluster 1 with 3 sector sites (1/3) at a soft blocking limit of L_{SB} = 34%: (K_{SB} = 1, N_{sec} = 3) implying 4 hopping frequencies (cf *Figure 8*). For the sake of simplicity, it has been assumed that the BCCH carrier is not included in the hopping sequence. Moreover, note that PC and DTX is not allowed on the BCCH carriers. With a DTX factor of 0.6 the effective soft blocking limit results in L_{SB} = 57%.

Thus, the spectral capacity in the 4/12 hard blocking scenario is 5.1 Erl/(site·MHz). This is almost trippled to 14.5 Erl/(site·MHz) in the 1/3 soft blocking scenario. These figures are depicted in *Figure 9* along with those for other reuse schemes and different available operator bandwidth.

To illustrate the capacity enhancement in a network exploiting all radio link options the spectral capacity of the tight reuse schemes is related to a

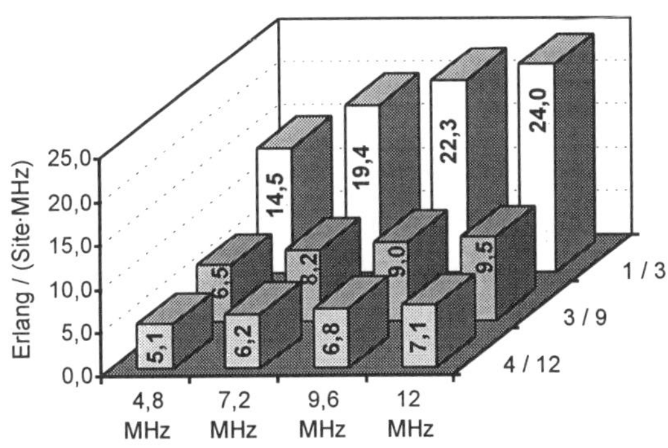

Figure 9. Spectral capacity for different operator bandwidth and different reuse schemes

conventional 4/12 reuse pattern. *Figure 10* shows the figures for the spectral capacity of the different reuse schemes normalized to the corresponding values of the reference reuse 4/12.

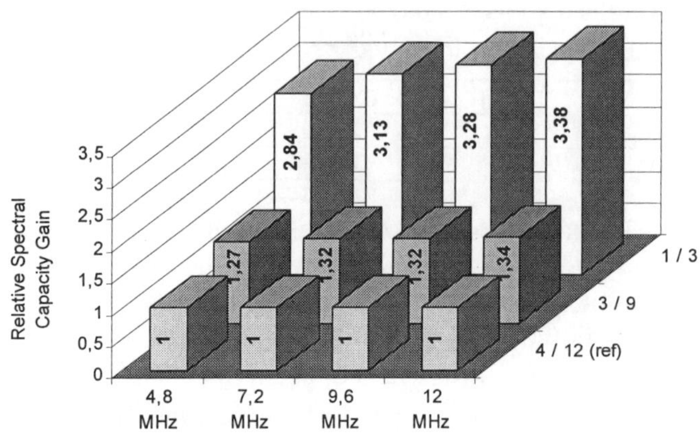

Figure 10. Relative spectral capacity gain of 1/3 and 3/9 reuse schemes compared to conventionally used 4/12 scheme

Evidently, a maximum gain in spectral capacity up to a factor of approximately 3.3 is achieved with the sectorized 1/3 clover-leaf cell reuse scheme. Even for operators with quite stringent bandwidth requirements it is possible to accommodate approximately 3 times the capacity of a conventionally operated network without radio link options.

8. CONCLUSIONS

The diversity effects of random FH are best exploited by slow moving mobiles in a dense urban environment. The gain from random FH increases as the number of hopping frequencies increases.

With an extremely low reuse factor ($K = 1$) there is a large number of TCHs and hopping frequencies available in each cell but the system load is limited by co-channel interference due to the small reuse distance. On the other hand frequency reuse schemes with large reuse distance ($K > 4$) are limited by the number of available channels per cell. Therefore the optimization of the spectral capacity is a trade-off between a large number of channels at a fractional load and a small number of channels at hard blocking limit. This is reflected in the figures for the spectral capacity η,

where our simulation results show that e.g. utilizing DTX with a total of 36 RF carriers (7.2 MHz) a fractionally loaded 1/3 reuse scheme provides a spectral capacity increase of about 200% compared to a non-hopping 4/12 reuse scheme at the same speech quality level.
We conclude that even with limited system spectrum, employing tight frequency reuse with random frequency hopping and DTX in combination with PC (quality based PC is a further option), cellular network operators are provided with enhanced spectral capacity to meet the demands of the rapidly growing subscriber penetration rates.
In addition to the radio link options, other techniques may be applied to further decrease the interference level within the network. The reduced interference level can directly be converted into additional capacity due to an increase of the soft blocking limit. Appropriate techniques to be applied are e.g. smart antennas or switched beam antennas which are being investigated in detail as well. It has to be noted that such techniques are best exploited if they are employed 'on top', i.e. in addition to the radio link options, since these capacity enhancement methods are the most cost efficient ones for the operator.

REFERENCES

[1] K. Ivanov et al, „ Frequency Hopping Spectral Capacity Enhancement of Cellular Networks", in Proceedings of ISSSTA96, 1996, pp 1267 - 1272
[2] J. Näslund et al, "An evolution of GSM", in Proceedings of the 44th IEEE Vehicular Technology Conference, 1994, pp. 348 - 352
[3] GSM 05.02, "Multiplexing and multiple access on the radio path", ETSI, 1996.
[4] GSM 05.03, "Channel coding", ETSI, 1996.
[5] GSM 05.05, "Radio transmission and reception", ETSI, 1996.
[6] GSM 03.30, "Network planning aspects", ETSI, 1996.
[7] H. Olofsson, J. Näslund, B. Ritzen and J. Sköld, "Interference diversity as means for increased capacity in GSM", in Proceedings of the 1st European Personal and Mobile Communications Conference, 1995, pp. 97-102.
[8] J. F. Withehead, "Signal Level Based Power Control for Co-Channel Interference Management", Proceedings 43th IEEE Vehicular Technology Conference, 1993, pp. 499 - 502

PART 3

IMPLEMENTATION ASPECTS

Chapter 1

GSM in the Indoor Business Environment

Howard Benn, Howard Thomas
Motorola GPD

Key words: GSM, indoor propagation, Phase 2+.

Abstract: As more business users use their GSM mobile phones in the office a big demand is being place upon the cellular operators. With limited frequency spectrum and high traffic requirements the basestations are moving indoors. Cell sizes are being reduced and high quality coverage is expected. This chapter presents a set of models that can be used to determine the coverage expected from an antenna in any given environment. With the antenna being closer to the user, new radio specifications are needed, these have been reflected in the new ETSI phase 2+ standard and are described in section 3. Solutions for transporting the RF to the antennas are presented in section 4.

1. INTRODUCTION

At a recent meeting I was given a business card from an associate, I gave it a quick look and just I was about to put it into my pocket I noticed that it only had a single phone number on it. After a brief discussion this turned out to be his mobile number, "We only have one phone now, I use in the car and in the office". The use of GSM phones in the office is one of the directions that the mobile market place has expanded into over the last few years, this chapter looks at that market, and the technical challenges it presents us with.

1.1 The Business Case

The award of new DCS1800 licences during 1996, the deregulation of many telecoms markets during 1997 and 1998 and the mandatory Liberalisation of Telecoms in Europe in 1998, provides the mobile market with the inevitable threats and opportunities accompanying any major change in environment. For existing operators, the threats come from new operators and the opportunities from the ability to offer new "converged" services and to reduce costs through new interconnect and backhaul agreements.

This changing environment is prompting operators to re-examine their businesses and to invest in new services that will maintain and grow them. In-building systems provide a platform for some of these services.

Any solutions for the corporate in-building market should offer:

- Business Opportunities for the mobile Operator
  - Subscriber retention through service differentiation
  - Subscriber retention through corporate "tie-in"
  - Opportunities for revenue generation through increased use
- A high level of service for the end user
  - Seamless service - the same functionality throughout the network
  - High speech quality
  - Security - call privacy and protection from fraud
- Low Operator cost of ownership
  - Rapid installation, simple operation
  - Smooth upgrade path
  - Maximum re-use of existing investment
  - Efficient and effective sales and distribution path

The advantages of a digital cellular based system include the attraction of capitalising on the 10's of millions of handsets already deployed.

To illustrate this, *Figure 1* shows the forecast installed base of GSM handsets over the next 4 years in medium and large businesses in Western Europe [1].

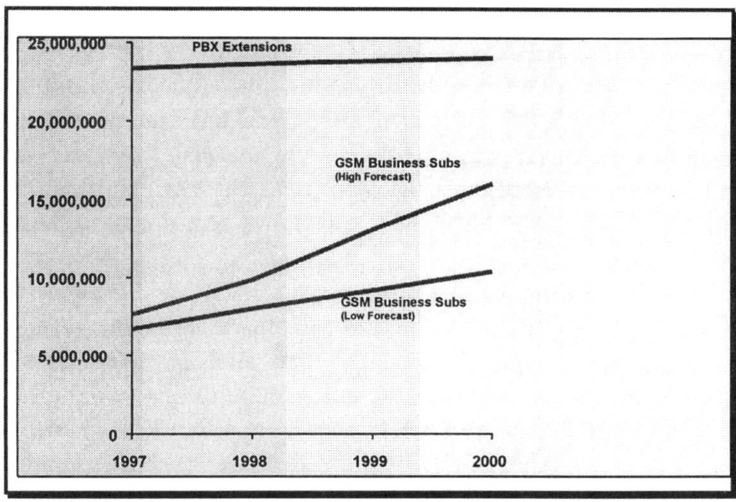

Figure 1. GSM Business Market

The number of PBX extensions in medium/large businesses in Western Europe is relatively stable over the next few years at approximately 24 million extensions. Even a pessimistic forecast of the number of GSM handsets used by business subscribers indicates that nearly 40% will have a mobile phone by the year 2000. This is a huge installed base into which operators can sell corporate in-building services.

2. SOLUTIONS FOR INDOOR COVERAGE

Coverage inside a building can be gained from 3 distinct methods, exterior illumination, indoor transmission networks and indoor BTSs. In any network there is a place for each of these solutions, but each comes with its own price in terms of cost, capacity, and quality.

2.1 Traffic Requirements

In some in-building systems it is the volume of traffic that will determine the type and size of system. Traffic can be measured as an absolute (total usage) or as an intensity (the amount of usage over a given time). Intensity is the measure used in traffic engineering, and its units are the Erlang (1 hour of usage per hour). The 'average' traffic usage of a business subscriber depends upon a great number of factors, but it is generally accepted as being in the region of 50mE to 300mE when the mobile is viewed as the primary phone.

A single GSM carrier with a combined BCCH has 7 traffic slots available, this can carry 3 E of traffic with a 2 % blocking probability. Due to trunking efficiency this increases to 9E for two carriers (15 traffic channels), and 15.8E for three (23 traffic channels).

If, for example, a building houses 360 people, the mobile penetration is 25%, and each user averages 50mE, then the total traffic is 4.5E. A two carrier system would provide sufficient capacity here. As the penetration increases to 50% and usage increases to 0.1E the total traffic becomes 18E. At this point a choice has to be made between a 4 carrier BTS, two 2 carrier units, or 6 single carrier units, the availability of spectrum becomes an important factor here.

2.2 Exterior Illumination

This is by far the most popular, and until recently, the only cost effective manner of giving users a degree of coverage within a building. Today's networks get inside by transmitting large amounts of power from outside. In some cases it is directed at particular buildings with directional antennas to increase the link budget. The penalty paid for this is that the reuse factor, or specifically the range, of the surrounding cells is reduced.

Major problems are encountered on the upper floors of high rise buildings as the C/I problem get worse, all the cells from a wide area around become interferers. It is common place for a mobile to show 5 bars but drop most of the calls.

At street level this problem is being addressed by the use of microcells. These, relatively low power, BTSs have very small cell radii, down to 50 m, increasing the network capacity while reducing the levels of interference they generate to other cells. The major factor that helps is the building loss. With microcellular antennas mounted below roof level, frequencies can be reused a relatively close distance away if buildings isolate them.

Having dedicated small cells within the buildings gets around this problem. The penetration loss works in our favour reducing the interference generated from outside and reducing any interference that is generated from inside.

2.3 Indoor Transmission Networks

This is a general category of system where the RF is transmitted from inside the building. There are several methods of obtaining this, ranging from repeating the exterior signals to having a dedicated BTS located within the building and using RF or fibre distribution systems.

2.3.1 RF Repeaters

When a building does not have any (or very poor) coverage then a repeater can be used to overcome the wall losses, receiving the exterior signals, amplifying them, then retransmitting .

Once the RF is available inside then one of the distribution systems described below could be used transport this around the building.

Within the industry, repeaters have been widely used to speed up roll out and provide coverage in remote areas. However they have to be used with great care to prevent problems with wideband noise, intermodulation, and stability. These problems have been described in the ETSI scenarios document GSM 05.50.

2.3.2 Coax distribution systems

To distribute the RF around a building passive coaxial transmission techniques can be used. These are most applicable to low capacity systems where a single, or multiple, carriers are taken from a single source and transmitted from multiple antennas at relatively low powers. This provides good internal levels while reducing the interference to the outside network.

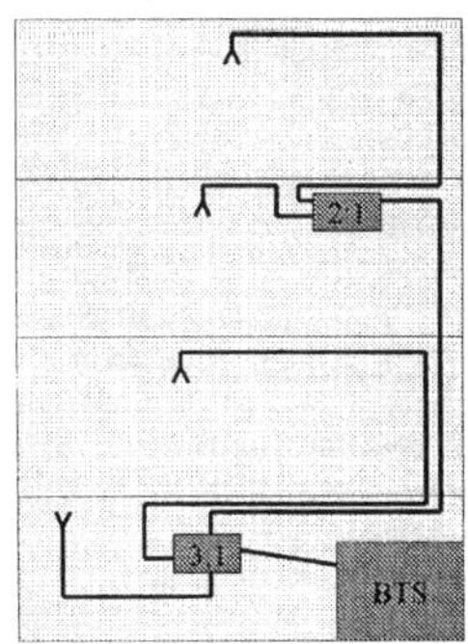

Figure 2. Coax Distribution System

Typically a conventional macro or microcellular BTS is located within an equipment room in the building, *Figure 2*. A splitter network and low loss coax cabling link several antennas to the single antenna port on the basestation. Because the dynamic range needed from each of the antennas is relatively low the losses introduced by the transmission network can be planned into the link budget calculations, *Table 1*.

Table 1. Calculating the link budget

BTS output power	+30 dBm	BTS sensitivity	-104 dBm
Cable loss	-2 dB		
3:1 splitter loss	-5 dB		
Cable loss	-5 dB		
2:1 splitter loss	-4 dB		
Cable loss	-4 dB		
Power at antenna	+10dBm	Sensitivity at antenna	-84 dBm

Initial installation of these systems can be problematic due to the problems involved in routing the coax cable. High quality low loss coax is made with a solid metal outer and a large diameter copper core, making it relatively thick (12 - 25 mm), and difficult to handle. It takes time to install and uses valuable ducting space.

Initial capacity expansion is relatively easy. If the system had a single carrier then capacity can be more than doubled by simply adding an additional carrier to the BTS. However this cannot be expanded without limit as most operators have a very limited frequency allocation, and the number of free carriers for use in the building may be low. In order to expand the process of cell splitting and reuse must take place.

2.3.3 Fibre distribution systems

These systems work on the same principle as the coax based ones but the RF is converted to a modulated light and transported around on optical fibre.

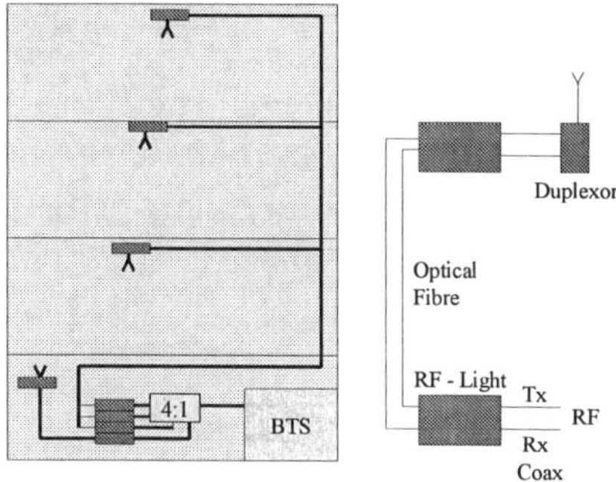

Figure 3. Fibre Distribution System

This has an advantage in installation, as fibre optic cable is far easier to handle however semi-skilled expertise is still required to terminate the fibre. Most systems use a single fibre per RF path (2 per antenna) and operate on a star rather then ring principle.

Expanding the number of carriers can be a problem as the RF to light conversion process is inherently non-linear. The output power per carrier is normally reduced to prevent any unwanted intermodulation products being generated, this will reduce the cell radius.

2.4 Distributed BTS Systems

This is a relatively new solution for high capacity buildings, and falls into the new ETSI BTS class P1. In order to increase the capacity of any cellular network, cells are made smaller; this principle is maintained within buildings.

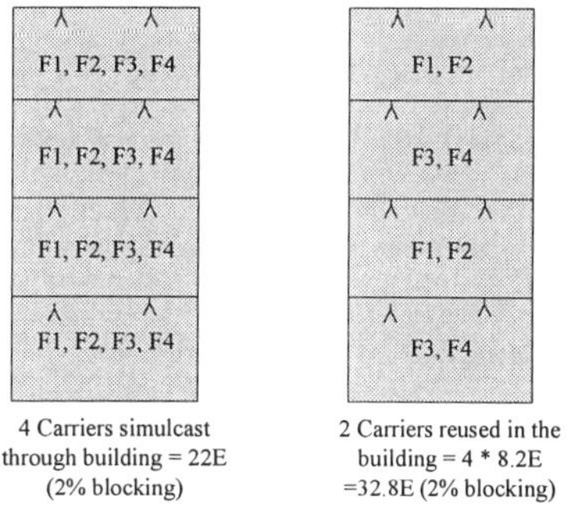

Figure 4. Expanding Capacity with Limited Spectrum

Most buildings can achieve excellent frequency reuse, especially between floors. There soon reaches a point, in which a distributed antenna system needs a single carrier per antenna. It then makes sense to place the radio with the antenna and to connect back to the controller in the digital domain. Using advanced communications standards such as HDSL [ETSI ETR 152] the wall units can be connected using the existing in-building telephony cabling. This is a massive step forward in reducing installation time and cost.

3. THE INDOOR PROPAGATION ENVIRONMENT

The indoor propagation environment is a complicated one. Path loss and minimum coupling loss are the fundamental parameters used in designing and planning indoor radio systems.

Path loss is the ‘normal’ parameter monitored in any propagation study. A number of path loss models that are applicable to the indoor environment are discussed, along with a comparison of measurements against those models for an example office.

Minimum coupling loss is small indoors because the mobile can be physically closer to the base than in either macro, or microcellular systems. The important factor derived from this is the percentage of time that an average user will cause, or be affected by, interference while close to the antenna.

Other parameters such as absolute time delay, and time dispersion are of interest, as they provide arguments for simplifying the GSM standards and reducing the complexity of in-building basestations.

Finally the effects of high levels of interference from the external cellular system are shown.

3.1 Path loss models

Studies by Motorola have shown, by comparison of experimental data with published models, that radio propagation in indoor environments may be modelled in quite a straight forward way to a reasonable degree of accuracy.

In particularly, in open plan offices, a model based on 0.4dB/m excess attenuation over free space can predict loss with a standard deviation of the order of 3dB. In more difficult environments with hard walled offices and lossy walls, details of the internal structure of the building must be included. A model based on the number of walls between the transmitter and receiver location combined with excess attenuation due to clutter can predict loss with a standard deviation of 7.5dB.

Due to the complexity of the indoor environment and the practical impossibility to describe all of the clutter within it, the accuracy of the statistical and deterministic approaches are very similar. As a consequence, the more complex approach represented by ray tracing is not justified for the indoor environment.

3.1.1 Summary of models in the literature

This section makes extensive use of several references, particularly, Hashimi [2], which provides a review of the different models available in the literature. The review identified four distinct path loss models which may be used for modelling indoor propagation in a variety environments to a reasonable degree of accuracy (~4-7dB RMS error).

The models described in the paper by Hashimi [2] are statistical and try to associate propagation loss with distance making an implicit allowance for obstructions. There is no attempt to describe the actual physical path the radio signal follows in these models.

It is possible to take a deterministic approach to modelling loss and ray-tracing is the prime example of this approach. The basis of these models is to use an accurate description of the indoor environment and trying to model the signal based on physical principles. However, these models require a much more complete description of the indoor environment than statistical models.

The models that have been found of most use were produced by Devasirvatham [3] and by Lafortatude [4].

The Devasirvatham [3] model is useful in open plan type offices where it permits very simple modelling of propagation using a single parameter to represent clutter. The Lafortatude [4] model is useful in the very heavily cluttered environment. Here, the details of the building cannot be ignored, as wall losses are too great. It allows a simplistic way to account for the details of the building by counting the numbers of walls in a way that would be easy to automate by computer. The model can also be used to model effects caused by diffraction, reflection, and penetration through multiple floors.

3.1.2 Clutter attenuation model

This model corresponds to a free space attenuation plus a linear loss (0.3-0.6 dB/m) and was validated in large and medium sized office and factory buildings in the range of 850 MHz, 1.9 GHz, 4 GHz and 5.8 GHz.

$$\text{Loss} = L_f - K\,d \qquad 0.3<K<0.6 \qquad (1)$$

$$L_f = 20 \log_{10} (\lambda / 4 \pi d) \qquad (2)$$

3.1.3 Building structure model

This predicts propagation losses by taking into account the variability of architectural configurations, building materials, and the effect of frequency.

This model associates logarithmic attenuation with various types of structures between the transmitter and receiver antennas. Adding these individual attenuations, the total path loss in dB can be calculated.

The prediction model requires a fairly detailed knowledge of the building configuration and associates a given communication link to the three propagation situations which, here, are denoted by the terms "transmission", "reflection" or "diffraction", respectively. Transmission would refer, for example, to the propagation losses due to obstacles; diffraction would refer to propagation around corners or adjacent corridors; and reflection to the signal gain which can be experienced when transmission and reception are taking place in the same room or corridor, for instance.

Two parameters, L_{ob} and G_{rm}, are introduced in the free space loss equation to represent the losses due to obstacles and the gain caused by multiple reflections.

The received signal power P_r will then be given by:

$$P_r = P_t + G_t + G_r + L_f + L_{ob} + G_{rm} \qquad (3)$$

where

P_r, P_t the received and transmitted powers (dBm)

G_r, G_t the receiver and transmitter antenna gains (dBi)

L_{ob} the loss due to obstacles (dB)

G_{rm} the gain due to reflections (dB)

and where the free space loss L_f is given by (2).

The objective is to estimate or model the values of L_{ob} and G_{rm} in a variety of conditions within buildings.

3.1.3.1 Transmission

Losses Through Walls: Isolating the transmission phenomenon, measurement sites have been chosen where the reflection and diffraction phenomena were considered negligible; that is, sites where there was no corridor parallel to the line joining transmitter and receiver. The resulting equation for the losses in excess of free space is:

$$L_{ob} = 3.7 - 1.5n - 10.7 \log_{10}(d) \quad (4)$$

where d the distance between transmitter and receiver (m), n the number of walls in the transmission path.

When the distance (d') between the transmitter and the first wall becomes larger than 4 m, the following correction factor should be added to (4)

$$\begin{array}{ll} 0 & \text{if } d' < 4\text{m.} \quad (5) \\ -7.8 + 15.3 \log_{10}(d') & \text{if } d' \geq 4\text{m.} \end{array}$$

Losses Through Floors: Typical attenuation values are 22 dB for one floor, an additional 16 dB (38 dB) for a second floor, and an additional 10 dB (48 dB) for a third floor. These measurements are representative of average signal strength levels over a five square meter area and refer to cases where the transmitting and receiving antennas were aligned in the vertical direction. Transmission losses between floors can be described then by the following equation:

$$L_{ob\ final} = L_{ob\ initial} - 27.5 - 41.7 \log_{10} p \quad (6)$$

where p is the number of floors and $L_{ob\ initial}$ is given by (4) and (5).

The effect of stairways has not been found significant except on the next floor, near the stairway, where additional signal levels in the order of 7 dB have been measured.

Other Cases: Other cases have also been studied; namely the effect of doors (when they are open), of windows, and of window screens. In the case of furniture, it has been found that the effect of low-level office furniture (chairs and desks) was not significant with antenna heights considered.

3.1.3.2 Reflection (corridor effect)

When a wave encounters an obstacle, there will be reflections, which can be channelled by certain wall configurations and which bring about significantly higher signal levels than would be the case with free space propagation. This effect has been observed in corridors and in large rooms when there was no obstruction between antennas. For the corridors without any obstruction the gain, G_{rm}, is given by

$$G_{rm} = 0.2 + 1.8 \log_{10}(d) \qquad (6)$$

3.1.3.3 Diffraction

This effect will generate signal levels higher than would be predicted by simple transmission. In theory, attenuation caused by sharp edge diffraction behaves according to the Huygens' principle and can be described using the Fresnel parameter ν. The computation of ν for each location of interest is somewhat tedious. Instead, a "geometric diffraction" parameter h, defined in *Figure 5* for the case of a lateral corridor, or for the case of a room adjacent to a main corridor where the signal source is located, has been used.

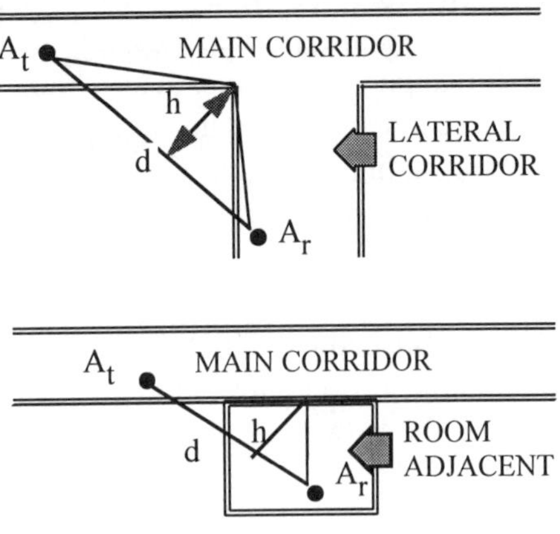

Figure 5. Diffraction Scenarios

For lateral corridors located more than 30 m. from the transmit antenna:

$$L_{ob} = -5.6 - 12 \log_{10} (h+1) \qquad d \geq 30 \text{ m.} \qquad (8)$$

3.1.3.4 Prediction accuracy

It is clear that the influence of different building materials and the great variability in architectural configurations limit the accuracy of any model and its applicability to a prediction method for signal attenuation within buildings. There will be a spread between the predicted and the real (measured) signal strength values. The probability density function of the estimation error shows that 68 percent of the measures fall within ± 2 dB of the estimated level and that the standard deviation is slightly lower than 3 dB. The accuracy decreases with the increase in distance, which of course corresponds to increasing uncertainty and diversity in architectural configuration, geometry, furniture, building materials, etc.

3.1.4 Power law model

According to this model the received signal power follows an inverse exponent law with the distance between the antennas;

$$P(d) = P_0 \, d^{-n} \qquad (9)$$

where P(d) is the power received at a distance d from the transmitter, and P_0 is the power at d=1 m. P_0 depends on the transmitter power, frequency, antenna heights and gains, etc. Path loss is therefore proportional to d^{-n} where n depends on the environment.

A great number of investigators have used this model. The reason is simplicity, on a logarithmic scale it gives a straight-line path loss with slope of 10n log d, and its previous successful application to the mobile channel.

Table 2 summarises published results on fitting indoor propagation data to an inverse exponential power law model. The table shows the environmental category, the power exponent and the standard deviation (where it is available).

3.1.5 Distance dependent power law

In this model the received power follows a d^{-n} law. The exponent n, however, changes with distance.

A distance-dependent exponent that increases from 2 to 12 with increasing d has been reported for indoor measurements carried out in a multi-storey office building. In the measurements the fixed transmitter was located in the middle of a corridor and the portable receiver was placed inside rooms along other corridors, on the same floor and on other floors. The value of n estimated from the data was 2 (for $1<d<10$ m.), 3 (for $10<d<20$m.), 6 (for $20<d<40$ m.), and 12 (for $d>40$ m.). The large values of n are probably due to an increase in the number of walls and partitions between the transmitter and receiver when d increases.

Table 2. Power Law Model Parameters

Environment	Exponent	Standard deviation
over all environments	1.2 - 6	16.3
factory floor	1.4 - 3.3	
factory obstructed	2.4 - 2.8	7.1
factory LOS	1.5 - 1.8	7.1
4 different buildings	1.81 - 5.22	4.3 - 13.3
office between floors	3.9	
office LOS	1.6	2.5
office non LOS	2.1 - 4.5	3.1 - 4.75
office hall ways	2	
office rooms	3	
University buildings	5 - 6	
Corridor	<2	
open plan shops	2	

3.1.6 Ray tracing

Ray tracing provides a deterministic way to estimate the received signal based on the geometry of the scattering environment. However, it is not a simple or cost free exercise to get a good description of the scattering environment, beyond, for example, a simple floor plan.

Several authors have compared ray tracing predictions with experimental results [5, 6]. The accuracy of the models, measured by the standard deviation of their predictions, is of the order of 4dB and is as much as 7.5dB into rooms. These levels of accuracy are not significantly better than the predictions of the statistical models.

The likely sources of inaccuracy are clutter in the environment, office furniture etc, which cannot realistically be included in the model. With respect to this, ray tracing can be no better than a statistical model.

Thus its current level of accuracy does not justify its use for in-building modelling in comparison with more simple statistical models.

3.1.7 An Example Office Environment

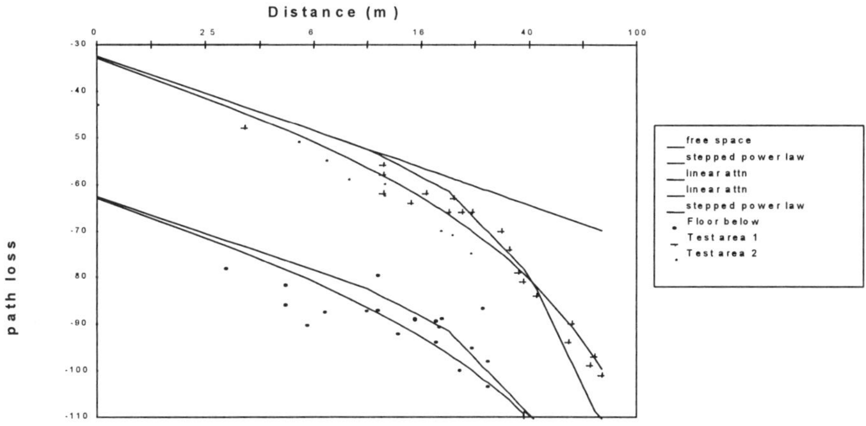

Figure 6. Example Office Environment

Figure 6 shows a number of the models highlighted above against one particular building. This case is a large open plan office with line of sight and large amounts of office clutter (2m partitions, furniture, etc.). Equation 1 with 0.44 dB/m clutter loss provides a good fit

3.2 Minimum Coupling Loss

Minimum coupling loss (MCL) is defined as being the smallest loss measured between the basestation antenna and the mobile. As a user moves around the building the propagation losses will change. ETSI therefore defined the MCL as a statistical parameter, being defined as the level that 0.1% of the propagation losses fall below. This is a critical parameter that determines the RF requirements for indoor systems, for example, wideband noise, blocking, and maximum transmit power requirements.

Following the approach used to set RF parameters for microcells, a measurement campaign was conducted by Motorola for ETSI to evaluate the MCL and its probability of occurrence in buildings.

The measurement site was on one floor of a 30 x 30 m office building. Three dummy base-sites were located in the office, operating at 900 MHz. The transmission sites used dipole antennas fixed to the ceiling (1.9m height) with 0 dBm into the antenna. A receiver, also with a dipole antenna, was moved evenly across the whole floor area, all together 30,000 power level measurements were taken. The probability density function of the data in presented in *Figure 7*.

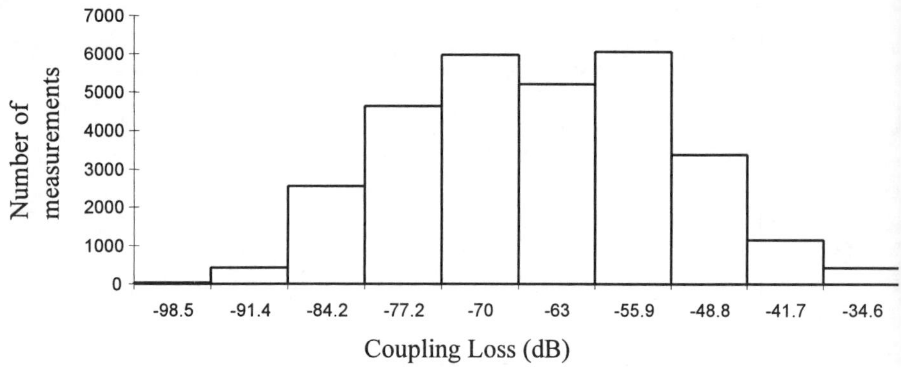

Figure 7. Distribution of Coupling Losses

Analysis of the data shows an MCL value of 34 dB at the 0.1% probability point. At 1800 MHz the measurements in the same environments gave an MCL of 40-42 dB. This ties in with the 34dB value measured at 900 MHz given the expected 6dB increase due to the halving in wavelength.

3.2.1 MS Margin (MSM)

This parameter is used by ETSI in their scenario analysis, section 4. It is used to relax the MCL from the measured 0.1% probability to 10%. A figure of 10% was seen to be reasonable as it assumes 10 handsets are within this distance (this originates from the multiple interference margin of 10 dB).

The following table gives selected values of the cumulative probability that the MCL will be smaller than the stated values.

Table 3. Cumulative Probability of MCL

Coupling value	Percentage of occurrence
<-33	0.03%
>-34	0.1%
>-36	0.53%
>-39	1.43%
>-42	2.86%
>-45	4.66%
>-49	9.58%

From the table the MCL at 10 % probability is -49 dB. When ETSI defined the microcellular classes in phase 2 a figure of 10 dB was used to go from the 0.1% to 10% probabilities. This has been maintained for the picocellular classes in phase 2+ and is used in the scenario analysis instead of the 16 dB that might be expected.

3.3 Delay Spread

In-building propagation is characterised by short, mostly line-of-sight paths between the MS and BTS. Consequently the delay spread between the paths is relatively short. Longer paths due to reflections from other sources (adjacent building for example) tend to be vastly reduced in amplitude due to the exit from, and re-entry into the building, doubling the penetration loss.

Table 4. Indoor Office Test Environment (Taken from table 1.2.2.1 Annex C, Region 1 REVAL validation group report)

Tap	Rel. Delay (uS)	Avg Power (dB)	Doppler
1	0.00	0	FLAT
2	0.07	-3.0	FLAT
3	0.17	-9.0	FLAT
4	0.31	-14.0	FLAT
5	0.41	-20.0	FLAT
6	0.53	-27.0	FLAT

Molkdar [7] presented a review of both wideband and narrow band experiments. In this review, results from various sources indicated worst-case delay spreads of 420ns with typical values of the order of 100ns. These values are far lower than the 16 micro seconds specified within GSM.

Exact measurements have been made by a large number of universities and commercial organisations, which have been fed into the COST 231 program. A new output from this work has been the FRAMES propagation study which will be used to 'test' various radio access schemes for UMTS. This happens to contain a delay spread model for in-building propagation, shown in *Table 4* below.

3.4 External Signal Levels

In all cellular communication systems C/I provides one of the limiting factors when determining capacity. The building penetration loss aids the internal systems by decreasing the external interference, hence enabling increased capacity.

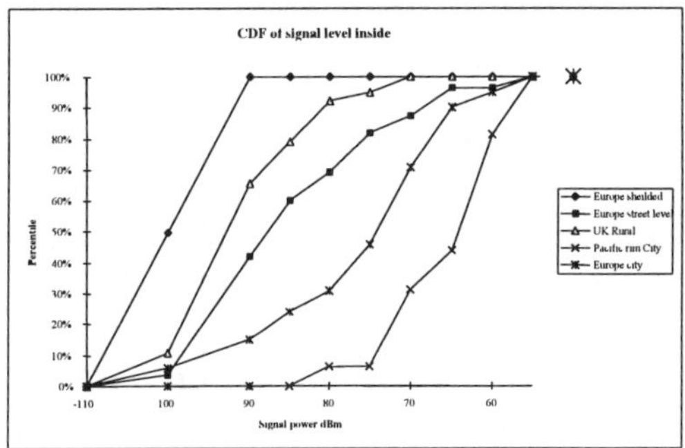

Figure 8. External Signal Levels

Figure 8 shows a wide variation of external signal levels measured from within buildings. When covered with lossy cladding (metalled glass, steel frames) and in the absence of any nearby cell sites the external network is virtually invisible. In some of the Pacific Rim countries, where the cell radii are down to 100m in the city centres, levels of -60 dBm are not unknown.

High levels from the exterior network can cause two problems. The first is to find available frequencies where the co-channel and adjacent channel interference does not degrade the GSM voice performance. In high rise buildings this is a particular problem as many of the surrounding cell sites are in line of sight, with little more than free space loss. Secondly if the call is to be originated on the in-building system then the mobile must be able to choose that cell when in idle mode. This process is defined in GSM 05.08 under the section entitled "Criteria for cell selection and reselection".

With phase 1 mobiles, most of the mobile population in the field today, the path loss criterion parameter C1 is used for cell reselection. This is dependent upon signal level and the system parameter RXLEV_ACCESS_MIN. Setting the building as a new location area provides additional control by setting the parameter, CELL_RESELECT_HYSTERESIS. In phase 2 an additional hysteresis factor, C2, was added to provide more control over the cell reselection. Controlling these parameters allows some margin where the exterior signal levels may be greater then the internal ones, and the call will still originate on the in-building system.

4. CALCULATING THE REQUIREMENTS FOR A NEW PICOCELL CLASS

ETSI defined a new phase 2+ work item in 1996 to study the effects of GSM indoors. This has resulted in a new BTS class - P1. The following section describes the reasoning behind the new parameters. The final decision on the actual levels to be used in the specification had not been made at the time of writing, the latest version of GSM 05.05 should always be consulted.

4.1 Background

4.1.1 Coordinated and Uncoordinated Users

Throughout the development of GSM it has been assumed that both mobiles and basestations can cover the complete GSM (or DCS) frequency band. It is also assumed that at least two operators will share that spectrum, and that they will be uncoordinated.

From a RF perspective the uncoordinated use leads to the toughest requirements. As an example in the UK there are 2 GSM 900 MHz operators Cellnet and Vodafone. When Cellnet installs an in-building system then no disruption to the Vodafone network will be tolerated. So if a Vodafone user is at maximum range from the external BTS, hence operating at sensitivity, the in-building BTS transmissions must be low enough not to affect that call.

4.1.2 Assumptions From GSM 05.05

For all the calculations hand-portable mobiles are assumed. This relates to MS class 4 at 900 MHz and MS class 1 at 1800 MHz. Since the output power is defined as peak power in the channel bandwidth, and the noise calculations are based upon average power in a smaller bandwidth a conversion factor is used. Conversion from peak power in 200 kHz to average power in 30 kHz is -8 dB.

The reference sensitivity performance is assumed under a C/(I + N) of 9 dB, where both C and (I + N) are average power measurements in the same bandwidth.

In order to simulate the effect of having multiple sources of noise and interference (many mobiles) an addition factor Multiple Interference Margin (MIM) is introduced into many of the calculations. This has been set to 10 dB. The MS margin (MSM) was covered in section 3.2.1.

4.2 Maximum BTS Output Power

The limiting factor in calculating the BTS output power is the level that can be transmitted without causing a problem to an uncoordinated user close to the antenna. The maximum output power from an in-building cell is

$$P = \text{MS blocking level} + \text{MCL} - \text{MIM} + \text{MSM} \qquad (10)$$

At 900 MHz this is +18 dBm, and at 1800 MHz +21 dBm

4.3 BTS Receiver Sensitivity

To match the up and down links with the maximum output power quoted above, the maximum receiver reference sensitivity at the BTS is

$$\text{BTS ref. sens.} = \text{MS output power} - \text{max. path loss} \qquad (11)$$
$$\text{max. path loss} = \text{BTS output power} - \text{MS ref. sens.} \qquad (12)$$

At 900 MHz this is -87 dBm and at 1800 MHz -91 dBm

However when an uncoordinated mobile is close to the in-building BTS the wideband noise will spread across the whole band. So there is a limit to the minimum signal level that a wanted mobile can go down to, this limit is determined by that external noise floor. There is little point in having a BTS with better sensitivity than can be used. The BTS receiver noise floor calculated using the uncoordinated mobile wideband noise at MCL is

BTS ref. sens. = MS wideband noise (in 200 kHz) - MCL + C/N (13)
MS wideband noise (in 200 kHz) = MS output power in 30 kHz - noise (dBc/100 kHz) + conversion factor (100 kHz -> 200 kHz) (14)

At 900 MHz this is -68 dBm, and at 1800 MHz -77 dBm

So there is a choice of receiver sensitivities based upon either a balanced link budget with the maximum cell radius or on the scenarios with an uncoordinated mobile at MCL. If the unit was designed with the maximum cell radius in mind, and an uncoordinated user came close to the in-building cell then the cell radius would decrease. Wanted users at the original cell limit would drop their calls.

To choose between them it is assumed that an operator will want the cell radius to stay constant under all conditions, but that the mobile should be operating at minimum output power. Using the original figures would achieve that goal. When a wanted mobile is at the cell boundary and there are no uncoordinated mobiles around, the wanted one can use power control to lower the output power, reducing interference and increasing battery life. At the cell boundary the mobile will be powered down to +14 dBm (25 mW) at 900 MHz and +16 dBm (30 mW) at 1800 MHz

4.4 BTS Spectrum due to modulation and wideband noise

The BTS wideband noise has to be reduced to a level which will not degrade receiver performance of an uncoordinated mobile at MCL.

Wideband noise = MS ref. sens. + MSM + C/N + MIM + MCL + conversion factor (200 kHz -> 100 kHz) (15)

At 900 MHz the wideband noise is -73 dBm and at 1800 MHz -65 dBm

4.5 Blocking Characteristics

The fundamental property of the radio being tested is the dynamic range. The upper limit is defined by the maximum power received from a mobile operating at MCL and the lower limit is the minimum signal level which must be received from a wanted mobile to meet the reference sensitivity requirement. In this scenario it is the wideband noise from the uncoordinated mobile that defines that lower limit. The highest level expected at the BTS receiver from an uncoordinated mobile will be

BTS blocking level = MS power - MCL (17)

At 900 MHz this is -1 dBm, and at 1800 MHz is -10 dBm

The minimum level of the wanted signal while the BTS is being blocked is calculated to be

BTS wanted signal during blocking = MS wideband noise in 200 kHz- MCL + C/N (18)

The actual level depends upon the frequency offset from the wanted mobile to the uncoordinated blocker.

Table 5. Minimum signal level with blocker present

Offset frequency	Wanted level at 900 MHz (dBm)	Wanted level at 1800 MHz (dBm)
0.6 - 1.6 MHz	-52	-61
1.6 - 3 MHz	-60	-65
< 3 MHz	-68	-77

Hence the dynamic range requirements are

Table 6. Dynamic Range

Offset Frequency	0.6 - 1.6 MHz	1.6 - 3 MHz	<3 MHz
At 900 MHz	51 dB	59 dB	67 dB
At 1800 MHz	51 dB	55 dB	67 dB

GSM 05.05 specifies the blocking in a different manner. Instead of leaving the blocker at the same level and changing the level of the wanted signal, it leaves the wanted signal at a fixed point (3 dB above sensitivity) and changes the level of the blocker. Maintaining the same dynamic range, a translation can be performed to present the figures in a similar format.

Table 7. GSM Defined Blocking Levels

Offset Frequency	0.6 - 1.6 MHz	1.6 - 3 MHz	<3 MHz
At 900 MHz	-38 dBm	-30 dBm	-22 dBm
At 1800 MHz	-38 dBm	-34 dBm	-22 dBm

4.6 Radio Frequency Tolerance

In the present system the mobile has to be designed to work with a Doppler shift caused by speeds up to 250 km/h at 900 MHz, and 130 km/h at 1800 MHz. This corresponds to a frequency offset of around 250 Hz in both cases.

Within a building the fastest a mobile would be expected to move at would be 10 km/h, corresponding to an offset of 10 Hz at 900 MHz, or 20 Hz at 1800 MHz. Therefore the absolute frequency tolerance can be reduced for the BTS.

At present the limit is 0.05 ppm (45 Hz at 900 MHz, 90 Hz at 1800 MHz). Taking the 1800 MHz case, the mobile can successfully decode signals with a 250 + 90 Hz offset at present = 340 Hz. The new requirement is (20 + frequency error) hence the new maximum frequency error is

$$\text{frequency error} = \text{present decode offset} - \text{new max. Doppler} \quad (16)$$

At 900 MHz the frequency error can be up to 0.32 ppm, or 0.18 ppm at 1800 MHz.

5. CONCLUSIONS

As more business users use their GSM mobile phones in the office a big demand is being placed upon the cellular operators. With limited frequency spectrum and high traffic requirements the basestations are moving indoors. Cell sizes are being reduced and high quality coverage is expected.

This chapter has presented a set of models that can be used to determine the coverage expected from an antenna in any given environment. With the antenna being closer to the user, new radio specifications are needed, these have been reflected in the new ETSI phase

2+ standard and described in section 3. Solutions for transporting the RF to the antennas were covered in section 4.

There seems little doubt that these systems will be highly successful. However the cellular supplies now must meet the challenge of providing small, low cost in-building BTS's that are easy to install and configure.

REFERENCES

1. Mobile Communications April 18th 1996
2. Hashimi, "The indoor radio propagation channel," Proc IEEE, Vol 81, No 7, July 1993
3. Devasirvatham, "Four frequency radio wave propagation measurements of the indoor environment in a large metropolitan commercial building," IEEE Globecom 1991
4. Lafortatude et al "Measurement and modeling of propagation losses in a building at 900 MHz," IEEE Trans VT vol 39, No 2, May 1990
5. Honcharenko, Burtoni et al, “Mechanism governing UHF propagation on single floor in modern office buildings”, IEEE Trans VT, Vol 41, No. 4, Nov 92,
6. Valenzuela, “A ray tracing approach to predicting indoor wireless transmission,” Proc IEEE VTC, May 1993
7. Molkdar, IEE Proceedings H, vol 138, No1, Feb 1991

Chapter 2

In-door Base Station Systems

Improving GSM to meet low-tier challenge

Marko I. Silventoinen
Nokia Research Center

Key words: GSM, low-tier, in-door, Home Base Station, Office Base Station.

Abstract: In this paper the base station issues for two indoor environments, home and office, are discussed. The two environments differ greatly in their characteristics and, consequently, different solutions are needed. At homes, the key issue is the cost of the base station. To meet the low-cost requirement, a MS based HW architecture for the base station is needed. The usage of only few timeslots instead of eight will cause the air-interface to change. Two proposals for air-interface standard are discussed: Adaptive Frequency Allocation (AFA) and Total Frequency Hopping (TFH) In offices, the key issue is spectral efficiency. The GSM operators will face the challenge of providing high capacity in urban down town areas with only few carriers. Contemporary base station fail in doing it in a cost effective way. New hybrid solution, where the best parts of a single cell and multicell architecture are used, is presented.

1. INTRODUCTION

In the contemporary and near future mobile communications systems the subscriber will be provided with the possibility of total mobility, i.e. of using their handsets in different environments including homes, offices, highways and shopping centres.

One strategy to achieve total mobility is to employ different radio communication systems in different environments, for example Digital Enhanced Cordless Telephone (DECT) at home and in the office and

Global System for Mobile Communications (GSM)[1] outside DECT coverage area or when GSM features are needed. However, in this way the subscriber needs either two handsets or an expensive dual-mode handset.

Another strategy is to improve GSM in such a way it can meet the low-tier requirements. Contemporary GSM products and standards do not perform too well in low-tier environments such as offices, residential areas, shopping malls and the like. Especially high system costs, doubts about the GSM ability to serve high traffic densities and the complexity of the radio network planning are key concerns. On the other hand, if the GSM could be improved to meet the low-tier requirements, it would be good for all. The economies of scale and fierce competition will drive both the infrastrucure and terminal costs down and this can be achieved best if there are not a multitude of different incompatible systems but only one. In this paper the possibilities to pursue this strategy are evaluated.

The penetration of the GSM is high, for example in Finland the penetration rate hit 30% recently and the other countries will follow. Among the persons working in offices the penetration rate is significantly higher The huge investments made to the Mobile Stations makes it attractive to develop new concepts, services and features while maintaining the air-interface untouched or only lightly modified.

The focus in this paper is on the base station. The base station is the key element in improving the GSM system radio interface while maintaining the compatibility with existing or slightly modified handsets. The network issues are not discussed in depth and the new GSM services such as GPRS and HSCSD has been left out completely.

In this paper the problematics of the two different in-door environments, home and office, will be analysed. In short, at home low capacity is to be offered by a very low cost device while in an office very high traffic demand must be served by limited number of carrier frequencies. In both environments the radio network planning should be avoided and the air-interface should remain untouched or only lightly modified. A couple of ways the GSM may meet the low-tier challenge in those environments will be presented and a rudimentary evaluation of the proposed methods will be given. Naturally the base stations can be used also in other environments: home base station at small offices and office base stations also in shopping malls, stadiums, supermarkets and even as a generic hot-spot solution.

[1] In this presentation GSM refers to any GSM based system, i.e. GSM 900, DCS 1800 and PCS 1900.

This chapter will continue with more profound analysis of the low-tier problematics. In chapter 2 the Home Base Station is presented and in chapter 3 the analysis of office BTS is given. The conclusion and the list of references will follow in chapters 4 and 5, respectively.

1.1 Low-tier Problematics

The current GSM base stations fit well in the environment where the traffic is moderate. In Figure 1 this area is left un-shaded. When the traffic is either very low or very high, the existing products are not feasible.

In Figure 1 the number of the sites and the traffic from the site are depicted. There are a great number of potential low traffic sites. For example, in the future when penetration rate hits 40% or so, it will mean that virtually every household has a MS and that each apartment or house can be regarded as a potential site. For those sites the smallest possible currently existing BTS configuration, 1 TRX BTS, with 7 traffic channels (TCH) clearly an overkill. One TRX BTS at home is simply too expensive and provides unnecessary high capacity and its employment will result in very high costs and poor spectral efficiency. That's why a low-cost, 1 or 2 timeslot Home Base Station is needed. Introduction of HBS will move the line where GSM can be used toward lower traffic sites as drawn in Figure 1.

Also, contemporary GSM BTS is not an ideal product for very high traffic environment, such as offices. The number of large offices is small, but the number of subscribers working in there is significant. The key consideration in the office environment is the spectral efficiency: In urban down town area the GSM operator has severe shortage of carrier frequencies and can not possible spent several of them to provide capacity for offices. For a GSM operator short of bandwidth the question is not in which way to provide capacity for offices but if it can do it in any way. By introducing an office-BTS that is more spectrum efficient than existing products, the line where GSM can be used can be moved toward higher density cell sites as drawn in Figure 1.

2. HOME BASE STATION

In this chapter the Home Base Station (HBS) concept, the fundamental problems every HBS system must solve some how and two proposed solutions are presented. First, the need for HBS is justified. Then, the overall system concept will be presented. HBS concept is not a clear one but multitude of rather different system architectures can be discussed under the term. In this presentation two major approaches, the cordless and base station approach are presented. The HBS as a concept has a set of unique problems concerning the air-interface that must be solved. Those fundamental problems are identified. There are at least two different proposal presented for GSM HBS: Adaptive Frequency Allocation scheme

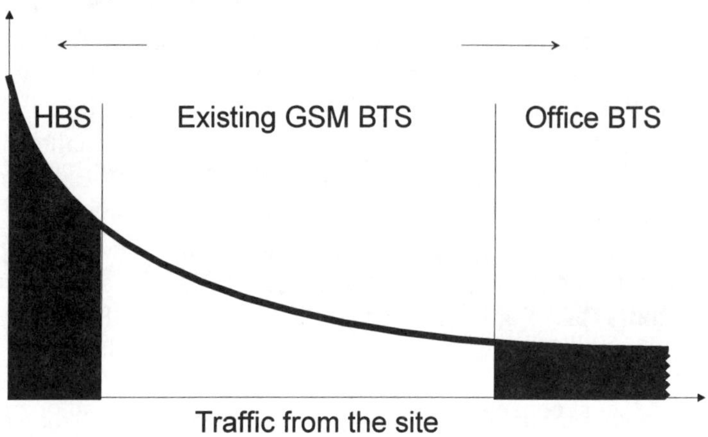

Figure 1. The traffic from a site, number of the sites and the different GSM BS feasibility

and Total Frequency Hopping scheme. These proposals will be shortly presented.

2.1 Need for Home Base Station

An important low-tier market segment is a residential use. In order to give a GSM subscriber incentive to replace her wireline telephone with a GSM terminal or at least encourage to use of the GSM, GSM operators may want to offer cheaper call rates for users when call is made at their home and to offer similar or better voice quality.

Within GSM, there are basically two ways to implement low-tier features. The other one is to provide users with a virtual low-tier service. That is, by adding more services in the GSM network, the user may be provided with the same services as a real low-tier network would do. In particular, in the domestic case, this would mean the use of Home Zones, i.e. predefined geographical areas where users are entitled to a lower tariff than outside that area. The Home Zone Calls (HZC) are already used in some networks. HZC is relatively easy to implement by network SW updates and does not require major additional investments from the operator.

The HZC concept has some drawbacks: With the current technology it is not possible to know the location of the user exactly. At the moment the serving cell is known and the call rates can be based on this information. In a contemporary cellular network the size of a cell is rather large, in the order of several kilometres in diameter, and the Home Zone becomes quite large, too. This will reduce the operators' revenue, since the HZC tariff will be applied in a much larger area than would be necessary for competing with wireline service. Another drawback is that a relatively expensive network infrastructure and spectrum would be used to provide the service which brings only marginal revenue.

The other solution is to provide GSM users with a possibility to invest to their own infrastructure. This can be achieved by Home Base Station (HBS), a low cost consumer device anyone can purchase, and operate. The HBS devices are currently available in the PHS and in the AMPS, for example, but not yet in the GSM. [Sil96a].

The benefits for subscriber are various: the congestion in the GSM network can be avoided, the battery life will increase due to GSM's power control ability to reduce transmission power when close to the base station, and the in-door coverage improves.

The HBS will bring benefits to the GSM operator, too: users get accustomed to their handsets and will eventually use it more also in the GSM network. This is important in emerging markets where customers must be taught to use mobile services. The subscriber participates in investing the infrastructure. The costly spectrum will be used, but the operator may bill the subscriber for the usage and since the HBS resides in-doors near the MS, the interference from the HBS to the macro network is modest. Finally, the operator does not have to worry about constructing in-door coverage but the subscriber unhappy with the perceived in-door service quality can be advised to purchase HBS. This is especially important in the early stages of the network build-up when the focus should be in offering a wide area coverage for out-door users and thus the in-door coverage may be poor.

2.2 Concept Overview

The HBS is a low-cost consumer product meant for the domestic environment and for small offices, i.e. for domains where the traffic density is low, the number of terminals is small and access rights are restricted to a small number of predefined MSs. It is not meant to be the solution for medium or large size offices where significant amount of traffic is generated or in public places like pedestrian areas or shopping malls where it is impossible to define beforehand which MSs are entitled to use of the HBS system.

The HBS can be connected to any PSTN where it uses the number allocated to that line by the PSTN operator. There should not be any minimal requirements concerned to the PSTN service level since they have a great variation even inside one country. From the PSTN's point of view the HBS is a normal terminal. In Figure 2 the HBS system overview is presented.

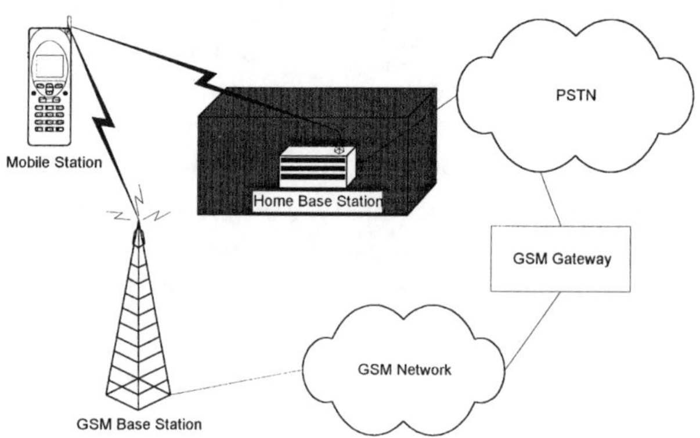

Figure 2: HBS system overview

2.2.1 Entering and Leaving Home Cell

The MS uses GSM network when it is not in the coverage area of a HBS it is allowed to use, i.e., when it is not in a Home Cell. When MS comes in a Home Cell, it will register itself, or camp, in the cell. A call transfer function in the GSM network will be activated to transfer calls to

HBS, i.e., to a PSTN number. For example, Follow-Me Diversion (FMD) can be used. The registration can be triggered automatically when the MS notices HBS's presence, by the user by pressing a key or periodically and it is essentially the same procedure as network reselection procedure in the GSM.

Only an authorised MS can registrate on HBS. The HBS will have a list of authorised MSs and will cross check, if a particular MS is allowed to registrate on it or not. The registrated MSs are able to see HBS either as a normal GSM BS or HBS service they are entitled to use depending on if the base station or cordless approach is chosen (see 2.3.). A non-registrated MS can not make any calls through HBS, with a possible exception of emergency calls. The user will be notified when the registration has taken place.

When a MS leaves a Home Cell, the call transfer function from the GSM to the HBS will be deactivated. The deregistration may take place when the MS makes the first location update in the GSM network or when the MS does not response the location update request or paging message sent by the HBS The user will be given a possibility to deregistate the MS manually. This feature is useful when the HBS has all the channels occupied and subscriber wants to make normal GSM call when camped to the HBS. Also, if the HBS is not able to provide full GSM services, the over riding of HBS is needed.

When a HBS is taken into use it has to be informed which MSs are authorised to use it. The introduction of a MS to the HBS, or initialisation, can take place over air or by inserting the SIM to the HBS, for example. Which ever way is used, the HBS will read the IMSI of the MS and add it to the list of authorised SIMs. The HBS will also write to the SIM the necessary access information like keys used in ciphering, HBS identification code, used frequencies and possible frequency hopping sequences.

2.2.2 HW implementation

The price of HBS should be low to be commercially attractive to the consumer, not significantly more than the price of a GSM handset. The key idea in HBS hardware implementation is to derive HBS from the existing MS by adding needed functionalities rather than trying to miniaturise current BTS. Since the power consumption, size or weight are not key issues for HBS, it can be manufactured using cheaper components than a MS. Also some expensive components such as display and battery can be left out. This will lead to a low-cost, low-power device where only some Time Slots (TS) out of eight available in a carrier are used.

It can be easily seen that in a normal case, where a house has only one fixed line connection, only one TCH is really needed and it is unnecessary to provide more capacity on the air interface. With ISDN connection and to support intercom calls[2], another TCH may be needed. Yet, in larger configurations, for example in small offices, even more TCHs can be useful. In any case, the most important thing to notice is that in a HBS there can be less than 8 TS in use. This has important reasons and consequences.

One of the reasons is that the use of only fraction of TSs will allow higher frequency efficiency. In theory, there can be as many as 8 HBS in one carrier, if each of them uses only one TS. Another reason is that when only few TS are used, a MS-like HW architecture is possible and hence the very low price level which is required for consumer product can be reached.

The main consequence is that the air-interface is no longer according to the GSM specifications. In the specifications, the GSM BTS is required to transmit continuously at a constant power level on the carrier frequency that has the BCCH, i.e. beacon frequency [GSM0508]. Since in the HBS all the TSs are not in use, the beacon frequency transmission is not constant and hence violates the specifications. The HBS air-interference must be re-designed, but in a way that minimised the cost effects to the terminals. In an ideal case, it would be possible to upgrade the MS to be HBS compatible by loading new SW.

2.2.3 Fundamental Problems

There are two major problems related to the radio interface every HBS concept must address somehow. These are the frequency allocation and inter-HBS synchronisation

2 The operator will have very limited means to introduce time-based billing scheme for intercom calls. The fact that the user have zero marginal costs of the airtime usage may increase the air-time usage dramatically. In general, to give a commodity for free encourages consumer to misuse the commodity. In mobile communications, completely free and unlimited billing schemes has been tried but abandoned. Intercom calls are a potential source of problems and its implementation is not only a matter of manufacturing costs, but also other factors must be carefully considered.

2.2.3.1 Frequency Allocation

Since the exact location of HBS is not known by the GSM operator and the location may change, it is difficult to allocate channels to HBSs prior to the actual power on. The number of HBSs in operation will be large, perhaps manifold compared to GSM BTSs. Also the possibilities for establishing efficient signalling link are restricted due to the fact the GSM operator has to essentially establish a call to the HBS via PSTN in order to transmit signalling data. Because of these three reasons, an autonomous channel allocation procedure must be used.

The Dynamic Channel Allocation (DCA) scheme is one possibility to overcome the frequency allocation problem. In the DCA a HBS scans the spectrum and selects a channel that offers low enough interference level. This requires receiving the down link (DL) signal from neighbouring HBSs as well as BTSs in the macro network. The downlink receiving is difficult to implement in HBS in a cost effective way. In GSM a Frequency Division Duplex (FDD) is adopted with link separation of 45 MHz. If a HBS is required to monitor a transmission band of another HBS, the bandwidth of the frequency synthesiser must be widened a lot and thus non-trivial extra cost is included. Without the possibility of monitoring the radio environment the HBS has no information about the channels at its disposal and hence it is unable to select a proper channel for communication.

2.2.3.2 HBS Synchronisation

HBS synchronisation is difficult to obtain, because HBSs do not have a common reference clock nor frequency reference. The PSTN does not provide HBSs with a synchronisation signal and the methods that can be used in the GSM BTS, such as high accuracy master frequency generators or exploiting Global Positioning System (GPS), are far too expensive for a low-cost product.

If the HBS synchronisation is not achieved it will cause problems both in frequency and time domain. The drift in the carrier frequency will sooner or later cause interference to adjacent carriers and may result in a situation where one carrier used by a HBS will distort two carriers. The drift in the time reference will cause the used timeslot to shift its place in the frame hence causing interference to adjacent timeslot. In short, without proper HBS time and frequency synchronisation, the spectral efficiency would be low.

2.3 HBS System Design Approaches

There are various different HBS scenarios presented. In this section I will describe two system design approaches, namely the cordless approach and base station approach. There are also other ways to implement HBS service, such as providing in-door coverage via RF signal repeaters, exploiting the cable-TV network to deliver signal from a BTS to a house, etc. These alternatives has not been discussed further.

2.3.1 Cordless approach

In cordless approach the HBS serves only as an access point to PSTN. The subscriber will be provided with PSTN services via HBS when at home cell. There is only minimal interaction between HBS and GSM network, Follow-Me Diversion (FMD) service being the major reason for any interaction, and thus the GSM-specific services would not be available at home. In practice this will mean that Short Messages are not delivered to the handset when at home and that the Supplementary Services will be different. The user will not be provided with seamless GSM coverage and she will have to be aware of whether she is in the GSM network or under a HBS. This approach is commonly called as Cordless Telephe System (CTS).

In the cordless approach, the HBS design will depend on in which network it is to be connected since the PSTN local loop specification varies from country to country. The variation will cause the HBS market to become fragmented and the product cost will increase. Since there is no clear specified interface between HBS and PSTN, it may happen that the vendors are not able to manufacture HBSs for every PSTN. This would reduce competition and further increase the price of the service.

The cordless approach will also make it quite difficult for a GSM-only operator to make revenue out of cordless HBS. The billing can not be based on air-time usage since it is out of control of the GSM-only operator. Instead of billing by minutes, monthly fees can be used, but, nevertheless, the GSM operator that operates also the fixed network will have a greater choice of charging methods and thus a clear competitive advantage. This will hinder free competition in the HBS market thus resulting to higher costs for consumers.[3]

[3] In this case the lacking capability of billing airtime usage does not necessarily lead to the misuse of airtime as in the case of intercom calls, since the user does not have zero marginal costs, but she hasto pay the fixed network tariff. However, in the cases there

The cordless HBS is clearly a partial solution for the problem. The benefits of this approach is that the standardisation effort is smaller and fast track product launch is possible. Yet another benefit is that the GSM network does not experience any modifications.

2.3.2 Base Station Approach

In base station approach the HBS functions as a part of the overlaying GSM network. Instead of being connected to BSC via A-bis interface as normal GSM BTS, the HBS is connected to BSC or MSC via PSTN network. This requires a modem for HBS and the standardised HBS-GSM network interface.

In this way all the GSM features can be provided at home and the seamless coverage is achieved. It may even be possible to provide a possibility for a handover between HBS and GSM network, a function which clearly improves the perceived service quality.

The call made using HBS as an accessdevice will be delivered to the destination via GSM network. This enables the GSM-only operator to monitor the usage of its spectrum accurately and introduce air-time usage based billing schemes. Since the GSM system will notice that a specific calls originates from a home cell, a lower tariff can be applied. The base station approach also enables the SIM-based billing schemes in contrast to PSTN-line based billing: the invoice can be directed to the actual user of the HBS, not necessarily to the owner of the PSTN-connection.

Since the HBS-GSM interface is standardised, every manufacturer may produce HBSs for every market area despite of the country specific PSTN standards.

The drawback of the Base Station approach is that it will take a relatively long time and a great effort to standardise the HBS-GSM interface. This will cause delays in HBS product and service introduction.

In Table 1 the relevant characteristics of the two approaches are summarised.

---

are zero fixed network tariff (calling to a toll-free number or during PSTN operators promotion campangies etc) similar problems may occur.

Table 1: Summary of the two HBS approaches

	Cordless	**Base station**
Services	PSTN	GSM
Air-time usage based billing	Not possible for GSM only operator	Possible
Object of billing	PSTN line	PSTN line or SIM
Market	Fragmented	Global
Changes to GSM network	No	Yes
Standardisation	Air-interface only	Both air- and network interfaces
Time of HBS availability	Relatively short	Relatively long

2.4 Air-interface Proposals

As explained in 2.2.2, the air-interface of the HBS will be re-designed. In this section I will describe two air-interface proposals, the Adaptive Frequency Allocation (AFA) and Total Frequency Hopping (TFH). It is clear that these two solutions are not the whole story. For example, for IS-136 system a DCA based air-interface concept has been published [Jar96]. For GSM new air-interface proposals may emerge, the proposals at design stage may be published and already published ones may be subject of improvements and refining.

2.4.1 Adaptive Frequency Allocation

The Adaptive Frequency Allocation method is described in [Haa97]. At the time of writing this paper, it is the most comprehensive proposal for GSM HBS air-interface specification and solves the two fundamental problems described in 2.2.3.

Despite of its name, it is not truly GSM compatible specification, but the system calls for some changes in air-interface. Allegedly, the modifications can be implemented with SW upgrades.

In the AFA concept, the frame structure is modified. Instead of 51-frame multiframe (51-MF), a 52-frame multiframe is used for the beacon carrier transmission. This will cause the HBS 52-MF to slide with respect to the 51-MF of the overlaying GSM network. This, together with the random jittering of the beacon signal from timeslot to timeslot inside the frame, enables MS to find HBS without interfering with normal cellular procedures. The period between HBS searches depends on the macro network environment the MS experiences. For example, when the MS notices that the CGI (Global Cell Identity) in the serving GSM cell is identical to that one it was able to listen during the initialisation with the HBS, the procedure is continuous. If the radio environment is different from what was experienced by the MS during the initialisation, the period can be 30 or 60 minutes.

The HBS system will be reserved one dedicated frequency, default BCH carrier, that is forbidden to use as a BCCH carrier by the macro network. This carrier is used when performing over-air initialisation of MS and HBS, when the interference situation in other carriers is uncertain or when the MS can not be accessed using other carriers. In normal operation, HBS uses other carriers it finds with the help of AFA algorithm.

In the AFA algorithm the key idea is to avoid interference from both macro network and other HBSs by selecting the least interfered channel for use. The GSM operator will first define a "Generic frequency list" that consists of the carriers the HBS system may operate. Using this list, the operator may restrict the use of certain carriers. Based on the long term interference measurements done by the HBS and by MSs registrated to the home cell, another list called "AFA candidate list" is generated. Using the MSs to measure the downlink interference level the HBS requirement for a wide band synthesiser is circumvented. If the HBS is able to measure downlink signal level, that information can be used, too. From "AFA candidate list" the best carriers are chosen for "Preferred BCH list", out of which the carriers used to transmit the beacon signal are selected.

The "AFA candidate list" is thus a list of least interfered carriers. In order to select best channels for communication, more lists are needed. First, a "DCS candidate list" is generated. In this list the interference situation of each channel of the carrier is stored, i.e. now the interference situation is examined at the accuracy of timeslot. From the "DCS candidate list", the best channels are selected for the "Preferred TCH list", from where the TCHs are taken.

To cope with the problem caused by a sudden drop of the TCH signal quality, for example when the reallocation of carrier frequencies in the macro network takes place, an escape handover procedure is designed. In a case of a radio channel failure both the HBS and MS are allowed to switch to a pre-defined escape channel without specific handover command.

The synchronisation problem is solved by letting a MS registered in the home cell to provide the HBS with the synchronisation signal it receives from overlaying GSM network. In a case of multitude of MSs registrated in the same home cell simultaneously, one MS is selected as the primary reference MS to avoid ambiguities in the case the other MSs receive signal from another BTS. The primary MS will indicate the HBS if has experienced a frequency or time offset. In the case there is no overlaying GSM network signal available, the inter-HBS synchronisation is not possible, but one may assume that in that case there is no shortage of free carriers, either.

The open question in the AFA is its performance. At the time of writing this paper (July 1997), no studies about the performance were publicly available. In a low-interference environment, such as residential areas, one may assume the AFA algorithm to work without major problems. In the case of high-interference environments, such as urban down town areas, the AFA algorithm faces a challenge to find a suitable channel.

2.4.2 Total Frequency Hopping

In the Total Frequency Hopping (TFH) a totally different approach was taken. Instead of trying to solve the fundamental problems presented in 2.3.3 , they were tried to circumvent.

In the TFH all the channels, i.e. the TCH as well as the signalling and broadcast channels transmit each burst in a different carrier frequency and the hopping takes place over all the available carriers the operator or the system has in a pseudo-random way. The hopping sequences for individual HBSs are independent of each other.

There are a few benefits in using THF in HBS concept:

1. The frequency hopping itself will increase the performance of the system because FH will help in fighting against the frequency selective fading. This is especially important when the speed of the MS is low and hence the fades last longer than when MS travels at high speed. Also, the interference diversity gain will improve the performance. [Olo96], [Höö96]
2. The TFH will eliminate the need for frequency planning since all the frequencies will be used in a pseudo-random manner and the additional interference to the overlaying system is evenly distributed over the whole spectrum in use.

3. There are no more problems with overlapping time slots or drifting carrier frequency due to the lack of HBS synchronisation. If the TSs of carrier frequencies overlap, a marginal average interference is generated but due to hopping the interference lasts only a short period of time.
4. The moderate TCH usage can be exploited by the system. If there is no traffic on a TCH, the interference is reduced and all the users will benefit from it. With a more rigid channel allocation system, an unused channel will still occupy bandwidth.

Practical implementation of TFH is considered to be rather straightforward. In the initialisation phase the frequency hopping sequence the HBS will use can be given to the MS. Hence, the MS will have a priori information about the TFH sequences of the HBSs it is allowed to use. With a priori information about the hopping sequence the synchronisation to the sequence is a trivial task.

There is a potential danger that a HBS system using the same frequency band as GSM will cause too much additional interference to the overlaying system. Obviously, if this is the case, the GSM operators would not allow the HBS system to use the same band. However, in [Sil96a] the simulation results reveal that the negative impact of the underlying HBS system to the GSM network performance is minimal.

On the other hand, the HBS system must itself work even though it suffers from the interference from high power overlaying GSM network side. In [Kuu96] the simulation result was that for reasonable TX power levels, the HBS system is more sensitive to the interference generated by the other network. In particular, the system parameters, such as wall attenuation and coefficients of the macro cell attenuation model, affect the system's feasibility. However, this study indicates that there are in general no problems in the service quality indoors. The possible problems arise if the HBS is indoors but the MS outdoors, e.g. in the garden.

2.5 Summary

In Table 2, the summary of the two proposed air-interfaces is given. The main difference between the two is that in AFA the interference is tried to avoid while in TFH it is averaged over the entire bandwidth.

Table 2: The summary of the two proposed HBS air-interfaces

	AFA	TFH
Interference	Avoided	Averaged
Frequency Hopping	Possible, requires several free carriers	Inherited part of the concept
Dynamic Channel Allocation	Inherited part of the concept	Not needed
Automatic HBS search	Yes	Not proposed
Stage of concept	Solid proposal exists	No clear proposal

3. OFFICE BASE STATION

In this chapter the office-BTS problematics is analysed. As in the previous chapter, we will start with describing why office-BTS is needed. The office radio environment characteristics are addressed only shortly. There are three major office architectures: single cell, multicell and a hybrid architecture. The hybrid architecture, named as In-door Base Station (IBS) will be described in more detail.

3.1 Need for Office Base Station

The main driving force in implementing a novel system for office environment is the lack of carrier frequencies in the urban areas. In Table 3, the bandwidths allocated to the ten largest European GSM operators are presented. It can be seen that in practice the operators has 25 - 60 carrier frequencies at their disposal. Assuming 18 reuse factor in macro cell network layer, the operators with narrow bandwidth have only 7 - 10 carriers for both micro cell and pico cell network layers, which causes severe problem in frequency planning.

The other important driver for novel office BTS is complicated network planning required by some contemporary architectures. The planning and managing of hundreds of multi-cell in-door systems is a formidable task and a source of extra costs for a GSM operator. Thus, ways of reducing the network planning and operation load are needed.

Table 3: European GSM operators. Bandwidth allocation, customer base and operating area. Situation 1.4. 1997. [Giu97]

Operator	Band (MHz)	Customers (millions)	Area (1000 km^2)
TIM	5.4	2.65	301
Mannesman	12.5	2.51	357
T-Mobil	12.5	2.45	357
France Tel. Mob.	12.5	1.557	544
Vodafone	8.2	1.478	244
Telefonica Mov.	8	1.345	506
Cellnet	8.2	1.125	244
SFR	12.5	1.014	544
Omnitel	5.4	0.913	301
Telia Mobile	5	0.865	450

3.2 Office Radio Environment Characteristics

The offices vary greatly in their RF characteristics. The material of the external and internal walls varies and, consequently, the penetration loss through the walls. The room layout differs from building to building, in a landscape office the internal walls may be few or non-existent. In some recently built buildings the attenuation due to ceilings may be small due to a several floor high open area in the middle of the building, for example This is why an office system must be designed in such a way it does not depend on any typical in-door characteristics (such as high attenuation between floors) but is able to cope with all of them (such as rapid field strength changes when entering the elevator, for example).

If an office system is able to cope with in-door radio characteristics without depending on them, it can be used also in other environments such as campus areas, shopping malls, super markets, stadiums and other sites, where traffic density is high. In its most generic form, the office-BTS could be used in any hot-spot area in the operator's network

3.3 Office-BTS Architectures

In this section, three office-BTS architectures are described. There are also other ways to provide offices with capacity and coverage, most importantly from an out-door cell. Those alternatives are not discussed in here.

3.3.1 Single Cell

In a single cell arhitecture an office consists of one single, large capacity cell. The signal distribution may be implemented by using leaky feeders, other forms of distributed antennas, or other known techniques.

At the moment with current BTS cost structure the single cell architecture is the most cost effective way to provide capacity. The large pool of TRXs is shared by the whole office thus resulting to trunking gain. The planning will become relatively simple since only the indoor-outdoor interference must be considered and there is no need for frequency planning inside the office building. Since the office consists of only one cell, there will be no intra-office handovers. This improves the perceived service quality and eliminates the tedious BSS parameter planning inside office.

The TRX usage rate is high due to the trunking gain. The single cell office system is able to offer this maximum capacity every where in the office, even in one spot, depending on the place of the traffic source. This kind of support for uneven traffic distribution will be useful for example outside a conference room during a break when people want to listen the voice mails and return the calls.

The drawback of the single cell system is that the capacity is limited due to poor spectral efficiency. The GSM operator needs to allocate the office BTS as many carriers as it has TRXs. Already at the moment and even more so in the future, there are GSM operators that simply do not have the required BW available. If they want to provide capacity in large offices they need other kind of BTS architectures, such as multicell architecture to be described in the next sub-section, hybrid scheme to be described in 3.3.3 and in more detail in 3.4. Or, they may use multi-mode or multiband handset, e.g. DECT/GSM or GSM/DCS handsets. The two latter alternatives are out of scope of this presentation

3.3.2 Multicell

In a multicell architecture there are a number of small capacity cells within one office building. Radio spectrum is divided between the cells and the frequencies can also be reused within the office in a similar way as in a traditional cellular network. The main characteristics of this approach are:

1. Network planning and managing are tedious, since the indoor frequency planning and the BSS parameter planning is needed.
2. Achievable capacity is high, but there will be no support for uneven traffic distribution.
3. Channel utilisation rate will be low.
4. Not very cost effective, because the costs due to planning, management and installation and low channel utilisation.
5. Intra-office handovers are frequent.
6. Frequency reuse is possible only when the number of cells exceeds the minimum reuse factor.

3.3.3 Hybrid Solution

In hybrid architecture we try to combine the good points of the two previous basic alternatives by creating one single logical cell and dividing it into a multiple smaller physical radio cells (or sub-cells). This concept has been named as IBS (In-building Base Station System) and it will be described in more detail in the next section.

3.4 In-door Base Station System

In the previous section, a hybrid architecture was introduced. In this section, the details of the novel architecture are given. This new concept is called In-building Base Station System (IBS) and it has been published in various papers, e.g. [Bro96], [Kro96] and [Sil97a]. Also, similar kind of systems are proposed. in [Lee91] and [Lea96], for example.

3.4.1 Architecture and Features

In the IBS all the TRXs are kept in one place, called as HUB, thus achieving trunking gain in signal processing capacity. When in traditional BTS each TRX must be given different carrier, in IBS it is possible use to use a multitude of TRXs with only few carriers. Hence, the signal processing capacity does not depend on carriers allocated to the system and high spectral efficiency can be achieved.

IBS system has number of low-cost RF-heads (RFH) around the building as drawn in Figure 3. These RF-heads will have a capability to radiate the signal of many carriers at the same time. Each RFH will form a physical cell, or radio cell, around it. The costs of forming this kind of cell are several order of magnitude less than with a traditional BTS.

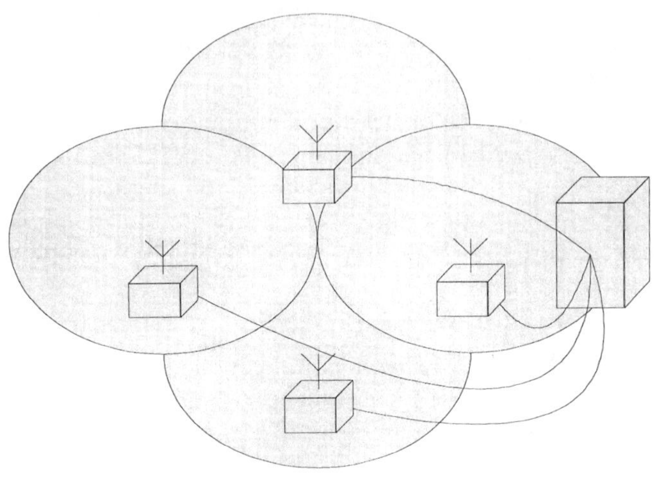

Figure 3: IBS system architecture

The carrier frequencies can be reused using the radio cells formed by RFHs. The same frequency and timeslot pair can be used in another RFH that is far enough in a similar manner as in traditional cellular network. The BCCH carrier is an exception: it will be transmitted via every RFH ensuring full compatibility with GSM specifications. In this way various improvements are achieved:

1) The air-interface remains as it is, i.e. existing handsets will be able to use IBS system without any modifications.

2) The size of the logical cell can be very large, in practice one cell for each office building. Because the call remains most likely in one cell, there is no need for inter-cell handovers during a call and this will make network planning a lot easier. The inter-system HO, i.e. the HO from or to office network, is trivial to implement.

3) The uneven traffic can be supported effectively. The signal processing power of the BTS is available everywhere and the limiting factor is the density of low cost radio frequency heads.

4) The frequency reuse inside a logical cell will increse the freuqency efficiency. In simulations presented in [Sil97b] it was shown that the IBS is two or three times more frequency efficient than a system created using single cell architecture.

The channel pool from which the channel is selected will be pre-allocated to the system, i.e. the network planner must tell the system which carriers the system is allowed to use. By this macro level cell planning a certain service level can be guaranteed in the office and the IBS does not cause uncontrollable interference to the overlaying GSM network.

3.4.2 Implementation

There are various alternatives on how to divide functions between RHF and HUB and how to transfer the signal from TRX to RFH and vice versa.

In a very simple case, the RFH is nothing more than an antenna and the signal from TRX to RFH is a radio frequency signal[4]. This configuration will allow fast product development and a low cost design, but the maximum distance between BTS and RFH is limited by the maximum output power of the TRX and the attenuation due to the cable between HUB and RFH.

Also other ways are possible. For example, if the signal from TRX to RFH is a base band signal, the existing Local Area Network of the office can be exploited. In this case, the RFH will include the functions required to convert the base band signal to radio frequency signal and vice versa.

No matter which kind of implementation is chosen, the IBS will not offer lower costs than a single cell system with distributed antennas because the additional units required in the BTS site. However, as single cell system requires as many carriers as TRXs its capacity is limited by the amount of BW available. By using IBS the operator can trade BW with extra HW: when the signal processing capacity of, let's say 10 TRXs is needed for a particular office but only 4 carriers are available, only IBS can be used, but it will cost more that normal system that could have been used if 10 carriers had been available. If the operator has a lot of BW or the traffic in the office is low, IBS is not needed.

3.4.3 Radio Resource Management

The problem with IBS system is that the channel definition will change, i.e. the channel is no longer defined as carrier frequency and timeslot but

[4] Note: The RF signal is kept in a cable, not transmitted to the air.

now also the place where frequency-timeslot pair is used, i.e. the RFH, has significance. This will cause the Radio Resource (RR) Management to be different from contemporary GSM BTS. The challenge in here is to make to modification in a way that no modifications to air-interface standards are needed.

A detailed description how the RR management can be arranged can be found in [Sil97a] In here a short summary will be given.

3.4.3.1 Basic Procedures

There are two basic produces from which more complex procedures can be derived. They are called RFH selection and Frequency-Timeslot selection.

In the RFH selection procedure the HUB finds the RFH that best serves the MS in question. The search can be accomplished by first measuring the received signal strength from all the RFHs, comparing the results and selecting the RFH that hears the MS best.

In the Frequency-Timeslot selection (F-TS selection) the BSC selects the Frequency-Timeslot combination that best serves the MS in question. The selection can be based on the idle channel interference measurements. The F-TS selection takes always place after the RFH selection, i.e. the system first selects the RFH and only after that measures interference situation in that specific RFH. In that way the BSC will be provided with the relevant information only and it does not have to be aware of the interference situation in any other RFH but the one that best serves the MS.

3.4.3.2 Call set-up

In the call set-up the system measures makes first the RFH selection based on the measurements done on the SDCCH channel used in call set-up. The time a MS spends on the SDCCH varies, but is typically in order of a couple of seconds. This gives the IBS enough time to perform the necessary measurements for RFH selection. After RFH selection, the F-TS selection takes place. The BTS will provide the BSC with the interference level measurements done via the selected RFH. The BSC then makes the decision which channel to allocate the call.

3.4.3.3 Sub-cell Handover

When an MS moves in the IBS system, another RFH comes closer and will eventually be able to serve the connection better than the current RFH. In this case a subcell HO is needed.

In sub-cell HO the RFH selection procedure is used. The HUB measures the signal level via the neighbouring RFHs and if better one is found, directs the signal to that RFH. This operation is an internal function of the HUB and it is invisible to BSC and MS.

3.4.3.4 Intracell handover

When two MSs have the same F-TS but at different RFH and they move closer to each other, the C/I of the communication link decreases. The BSC notices this by observing in a normal way the measurement reports it receives from both MSs and the HUB. The received signal level (RXLEV) remains high, but the received signal quality (RXQUAL) gets bad so BSC commands a normal intracell-HO. This procedure is essentially the same as F-TS selection: the BSC selects best F-TS combination to be used in the specific RFH

3.4.3.5 HO from IBS

This HO takes places in a normal manner, without any changes. When the MS moves away from the office network and the no better RFH can be found, the BSC experiences drop in RXLEV. The BSC commands MS to TX at higher power level, but it will help only temporarily. Eventually BSC will get measurement reports from MS indicating that signal level from an external cell is better by a certain margin. This will trigger a normal inter cell HO from the IBS network to the external GSM network.

3.4.3.6 HO to IBS

The problem with the HO to the IBS system is the RFH selection. The best RFH must be known before the BSC can make the F-TS selection for the incoming MS.

When MS is handed over to the IBS it will be allocated a TCH on the BCCH carrier. Since the BCCH carrier is transmitted via every RFH, the RFH selection can be deferred and performed later using similar procedure as in the intracell HO.

3.5 Summary

In this section the three different office system architectures were discussed. In Table 4, the summary of the office architectures is given

Table 4: Summary of office arcitectures

	Single cell	Hybrid	Multicell
Number of logical cells	1	1	Many
Number of radio cells	1	Many	Many
Handover inside office	No	Sub-cell HO, not visible to MS or BSC	Yes
Freq. planning Param. planning	Indoor-outdoor	Indoor-outdoor	Indoor-out-door, in-door indoor
Freq. efficiency	No reuse => poor	Reuse => good	Reuse => good
Support for uneven traffic	Yes	Yes	No
TRX utilisation rate	High	High	Low
Availability	Yes	Not yet	Yes

4. CONCLUSION

In this paper the problematics of in-door GSM systems is discussed with the focus on the base station. Two different in-door environment, called as Home and Office, are identified and their characteristics are summarised in Table 5.

For Home environment the key issue is low cost device. This is achieved using a MS based HW, which results to into modifications to the air-interface. The HBS is a private device with restricted access rights and supports only few terminals. There is no absolute need for either inter-HBS handovers or handover from HBS to GSM or vice versa. However, especially the HO from HBS to GSM would be nice to have

Table 5: In-door environment summary

Environment Characteristics	Home	Office
Modifications to GSM air-interface	Minor	None
Pre-determined access rights	Yes	Possible
Traffic volume	Low	Very high
Operator control	Via PSTN	Normal
Frequency planning	Impossible	Possible but tedious
Number of MSs in the system	Low	High
Nature of the BS	Consumer product, autonomous	Traditional telecom network product
Inter system HO	Nice-to-have, May be possible	Required, possible
Intra system HO	May be needed in special occasions, possible	Required, possible.
Other operating environments	Small offices	Shopping malls, campuses, out-door hot-spots
Key issue	Low cost BS	Frequency efficiency

The characteristics of the HBS are quite different from the base station required in the offices, or other high traffic density places, where a great number of subscribers must be served. In offices the access rights may be un-restricted, the number of MSs in the system is very large and the handovers inside an office system as well as to the GSM network are clearly a must. The key issue is to provide the in-door capacity in much more frequency efficient way as is done nowadays. A hybrid solution where the best parts of a single cell and a multicell architectures are combined may be the architecture to be used in officies.

As can be noticed in this paper, the GSM has a multitude of ways to grow to meet the low-tier challenge. In the HBS context, some modifications for the air-interface specifications are needed, but in the offices a significant performance improvement is possible by just redesigning the BTS.

REFERENCES

Bro96 Broddner S., Lilliestrale M. and Wallstedt K., "Evolution of Cellular Technology for Indoor Coverage", 11th International Symposium on Subscriber Loops & Services, Feb. 4 - 9., 1996, Melbourne, Australia

Giu97 Pietro Porzio Giusto, "Achieving the Optimum Balance of Capacity, Quality and Cost-Efficiency in Planning for Maximimum Capacity", presentation in Global Capacity Forum '97, June 24-25 1997, London, UK

GSM0508 European digital cellular telecommunication system (Phase 2); Radio subsystem link control (GSM 05.08), version 4.8.0, 21 Jan 1994, ETSI

Haa97 J.C. Haartsen, "Air-interface specification for GSM-compatible personal radio communications system", version B Jan 22. 1997

Höö96 Höök M., Johansson C. and Olofsson H., "Frequency Diversity Gain in Indoor GSM Systems", 46th Vehicular Technology Conference, Atlanta, USA, 1996

Jar96 Jarrett et al, "System Design for an Autonomous IS-136 Personal Base Station", 46th Vehicular Technology Conference, Atlanta, USA, 1996

Kro96 Kronestedt F., Frodigh M. and Wallstedt K., "Radio Network Performance for Indoor Cellular Systems, 5th IEEE International Conference on Universal Personal Communications, Sep. 29 - Oct. 2, 1996, Cambridge, USA

Kuu97 Maija Kuusela, Marko I. Silventoinen Mika Raitola and Pekka Ranta "Feasibility for a GSM Private Cordless Base Station Based on Total Frequency Hopping", Workshop on Multiaccess, Mobility and Teletraffic for Personal Communications, May 20-22 1996, Paris, France

Lea96 Chin-Tau Lea, "A new network arcitecture for wireless ommunications", 7th IEEE International Symposium of Personal, Indoor and Mobile Radio Communications, Taiwan

Lee91 William C. Y. Lee, "Smaller Cells for Greater Performance", IEEE Communications Magazine, Nov 1991

Olo96 Olofsson H., Näslund J. and Sköld J., "Interference Diversity Gain in Frequency Hopping GSM", 45th Vehicular Technology Conference, Chicago, USA, 1996

Sil96 Marko I Silventoinen, Maija Kuusela and Pekka Ranta: "Frequency Hopping HBS", IEEE International Conference on Personal Wireless Communication Feb. 19-21, 1996 New Delhi, India

Sil97a Marko I. Silventoinen and Harri Posti "Radio Resource Management in a Novel Indoor GSM Base Station System", The 8th IEEE International

Symposium of Personal, Indoor and Mobile Radio Communications, Sep. 1-4, 1997 Helsinki, Finland

Sil97b Marko I. Silventoinen, Petri Patronen, Jari Ryynänen and Eija K. Saario, "The Performance of Novel Indoor GSM Base Station System", to be presented in Multiaccess, Mobility and Teletrffic in Personal Communications 1997 workshop, Dec. 15-17, 1997, Melbourne, Australia

Chapter 3

GSM/Satellite Issues

HEIKKI EINOLA
Nokia Telecommunications

Key words: GSM, Service extension, Integration, S-PCN, GSM MAP protocol, GSM SIM, GSM SIM-ME interface.

Abstract: In this presentation, the mobile satellite service (MSS) is regarded as complementary to the corresponding terrestrial systems. The MSS may offer an outdoor service extension to the terrestrial mobile systems. This can be achieved with appropriate integration of the satellites with the terrestrial system. This chapter considers the integration of mobile satellites into the GSM system. On one hand, the services available and being planned for GSM place many demanding requirements in the satellite systems aiming to integrate with GSM. On the other hand, GSM provides a good core network platform for integration as it is a modular system supporting full-fledged international mobility.

1. INTRODUCTION

Satellite communications is a multibillion dollar industry with world-wide markets. Satellites allow the communications to reach practically any outdoor location on the surface of the Earth with no terrestrial transmission lines. The application of satellites includes the fixed satellite service (FSS), which may involve the extension of terrestrial transmission lines. Another application is the direct broadcasting satellite service (DBSS), which can be utilised for broadcasting television to receivers distributed over a large geographical area. Yet another application is the mobile satellite service

(MSS) that is characterised by mobile terminals being connected directly to the satellite instead of to a terrestrial basestation.

In this chapter, the MSS is regarded as to complement terrestrial mobile communications. The potential dimension provided by the satellites is in their ability to offer communication services into larger outdoor areas than the corresponding terrestrial mobile systems. In this scenario, the terrestrial mobile system is the preferable way to communicate and the satellite system would only be accessed when the terrestrial system is unavailable. However, one can also envisage that MSS would be used as the primary access instead of the terrestrial.

So far, the mobile satellite systems have been characterised by the use of geostationary (GSO) satellites, which remain fixed in place relative to the surface of the Earth. These services have typically been utilised by a small number of professionals. In the near future, however, many new satellite systems will be introduced. The proposals are generally referred to as global mobile personal communications by satellite (GMPCS). A number of these systems will rely on non-geostationary satellites that are closer to the Earth's surface than those in the GSO. The utilisation of large number of non-geostationary satellites represents a revolution away from the utilisation of a single or a relatively small number of geostationary satellites in the systems.

Certain emerging systems are relevant from the GSM point of view as they aim at providing services comparable to GSM with single subscription into a handheld terminal. Moreover, they aim at utilising existing GSM platforms and protocols. These systems can be called either satellite personal communications networks (S-PCN) or satellite personal communications services (S-PCS) systems. S-PCN can be briefly defined as a network offering personal communications services for individual users into handheld terminals via satellites. Personal communications itself includes ideas on communications with a single subscription anywhere, anytime at a cost effective price into a personal handheld terminal.

The purpose of this chapter is to consider satellite issues in relation to GSM. The mobile satellite systems are considered in three different ways, which include engineering, service and integration point of view. The engineering point of view places emphasis on the satellite system that represents a technological challenge requiring competencies on many branches of physics and engineering. The service approach considers how satellites could be utilised to complement terrestrial mobile systems. Finally, adaptation of the existing GSM platforms for the satellite systems is considered. These aspects can be characterised with integration of satellites into terrestrial systems. This text is organised as described above, the emphasis being on the integration of satellites into GSM.

2. MOBILE SATELLITE SERVICE SYSTEMS

2.1 Introductory remarks

Satellite communications can be treated as an extension to the terrestrial line-of-sight (LOS) transmission. The key requirement in LOS is that the transmitting and receiving antennas see each other in terms of direct propagation of radiowaves. Any shadowing, e.g., caused by the vegetation or buildings in between the antennas is undesirable especially in satellite communications. This is because the attenuation caused by the great distance is significant, and as a consequence, the received signal levels are low. As a result of this, satellite systems can generally provide service in the outdoor environment only.

2.2 General system description

A mobile satellite system can be regarded as a modular system comprising several rather independent components that are typically supplied by different manufacturers. *Figure 1* is a simplified presentation of such a system. The system contains both a terrestrial and a satellite component. The terrestrial component includes the core network (CN), which provides services and switching for the system. It also provides connections to the external terrestrial networks. An example of a core network is the GSM network subsystem (NSS), which is composed of the mobile switching centre (MSC) and the registers, such as the home location register (HLR) and the visitors location register (VLR). The satellite component in the system appears as a radio access network (RAN) to the core network. In a modular system the core network does not need any information about the internal structure of the radio access network. As a result, the requirements on the core network are mostly independent of the radio access.

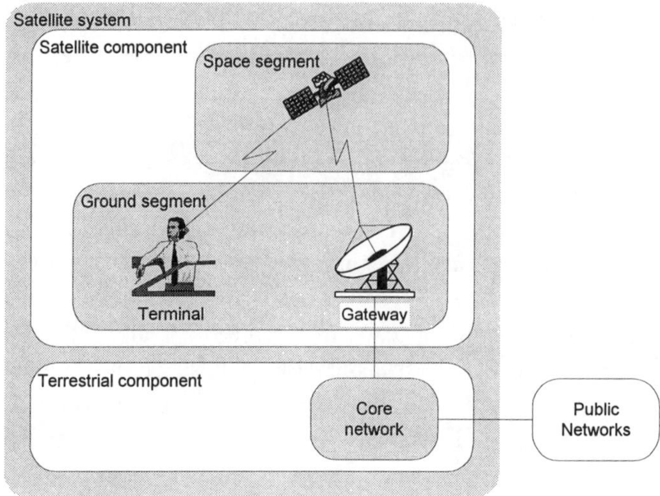

Figure 1. Satellite system components. Also, the system needs facilities to control and monitor the satellites, which are not included in this illustration.

The satellite component comprises a ground and a space segment. An overview of the satellite component can be found, e.g., in (Pritchard et al, 1993). The ground segment consists of earth stations capable of communicating with the satellite. Two types of earth stations are distinguished between here, namely mobile terminals and gateways. The terminals are used by the subscribers, whereas the gateways are part of the radio access network. The space segment consists of satellites that appear as repeaters for the communication system in a sense that no end user communication neither originates nor terminates at the satellite.

Satellite orbit altitude selection is one important step in satellite system design. The orbits that can be utilised in practice can be classified according to the orbit altitude and dynamics. Satellites orbiting in close to circular low Earth orbit (LEO) and medium Earth orbit (MEO) are moving relatively to the surface of the Earth. The movement implies that more than one satellites are needed in such a system to offer continuous service even into a single fixed location on the ground. Satellites in the geostationary orbit (GSO) remain in fixed positions relatively to the surface of the Earth. This implies that geostationary satellites are very suitable to provide regional or fixed service. A geostationary satellite may have a service area of roughly one third of the Earth's surface. Yet another class of orbits consists of highly elliptical orbit (HEO) (Pritchard et al, 1993), which is a non-geostationary but elliptic orbit.

Table 1 presents important features of different types of satellite systems. Advantages of systems with LEO or MEO satellites are the shorter delay in communications and lower attenuation when compared to systems with more distant geostationary satellites. However, the movement and the larger number of satellites makes the system more complex. For example, the movement of the satellites induces handovers even if the subscriber was not moving at all. The complexity increases risks related to the system. The risks of non-geostationary systems are discussed, e.g., in (Gaffney et al, 1996).

Table 1. Summary of technical characteristics[1]. Data in the table are adapted from (ETR 12.02) and (Brugel, 1994).

Characteristic	GSO	HEO	MEO	LEO
orbit altitude in km	35786	perigee: 1000[2] apogee: 39400	≈ 10000	≈ 700-1000
propagation delay in ms (mobile-satellite-Earth)	280	200-310	80-120	20-60
orbit period in hours	24	8-24	≈ 6	≈ 1.5
satellite handover during a call	unlikely	every 4 - 8 hours	every 2 hours	every 10 minutes
number of satellites for near global cover	3-4	5 - 12	10 - 15	> 48
operational complexity	simple	moderate	moderate	high

2.3 Mobile satellite services complementing terrestrial mobile services

Mobile satellite communications are here regarded as complementary to terrestrial mobile communications. Basically, this is due to the lower capacity offered by the mobile satellite systems. Also, the cost of using the satellite is typically higher. As a result, the usage of satellites in relation to terrestrial systems is justified when terrestrial systems are either not competitive, not applicable, or less developed or not developed at all (Ananasso, 1996). In practice, this means that mobile users would utilise mobile satellite access only when the terrestrial mobile access cannot be utilised.

[1] All values are for guidance only.

[2] The orbit altitude values correspond to the so-called Molniya orbit.

The potential of satellites to complement terrestrial systems is dependent on the properties of the existing wireline and wireless system infrastructure together with the area characteristics. Mobile satellite systems offer service into large outdoor areas. Therefore, the satellite systems may offer extension for terrestrial mobile systems both at regional and global level. *Figure 2* illustrates different environments in terms of terrestrial mobile system coverage. The possible utilisation of satellites in these environments as a service extension to terrestrial mobile systems is discussed below.

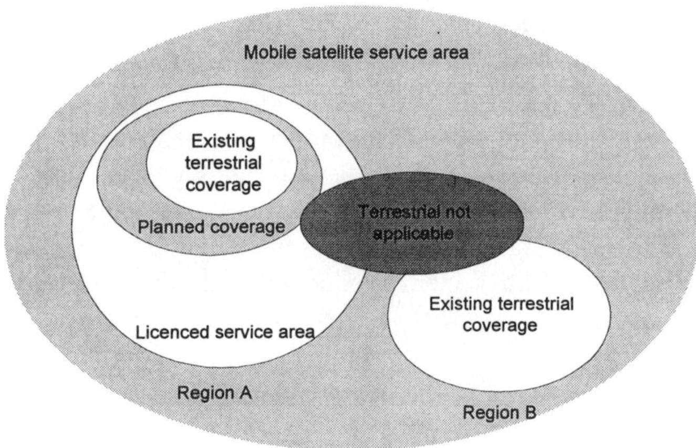

Figure 2. Environment types.

The use of satellites to complement terrestrial mobile systems includes extending the service into areas, where the terrestrial system cannot be accessed either due to economical or incompatibility reasons, or because the terrestrial system is not applicable. The extension may also include utilising the mobile satellite service as an interim way to offer comparable service until terrestrial system reaches the planned coverage (*Figure 2*). In this case, the satellite system has to be in operation prior the terrestrial system. Typically, areas of low traffic within the licensed service area (*Figure 2*) may be covered only after significant delay. Moreover, certain areas with very low traffic may not be covered at all, due to economical reasons. These areas falling totally outside the planned coverage may include, e.g., highly rural environments, wilderness, and deserts.

The extension due to incompatibility reasons is related to the lacking of common standards, or commercial agreements between terrestrial operators in different regions (A and B in the *Figure 2*). Therefore, the area in question may comprise a large geographical region (e.g. a large country or a continent). As a result, there is no single terrestrial system that could provide service in the whole region or even in both regions. It is also possible that the same terrestrial system is used in both regions, but the system does not support mobility between networks. Finally, the service extension into environments, where terrestrial systems are not applicable (*Figure 2*), may include for example the oceans. In these environments, a satellite system will be the only possible way of communication.

3. OVERVIEW ON INTEGRATION

Integration allows some form of co-operation between two different systems, which results in cost savings and service benefits. The cost reduction is based on utilisation of the same technologies or equipment for both systems. Service benefits mean that a user would receive and perceive the services in a similar fashion in both systems. Integration is also a way to describe how the satellite system may complement the terrestrial mobile system as defined in section 2.3.

3.1 The four levels of integration

In the integration of terrestrial and satellite systems there are a number of relations and approaches to be considered. Several levels of integration can be defined so that each level contains the basic features of the previous one. *Figure 3* illustrates the levels as defined in (ETR 093) and the most essential property of each level beyond geographical integration. In the following, main characteristics of each level are briefly described.

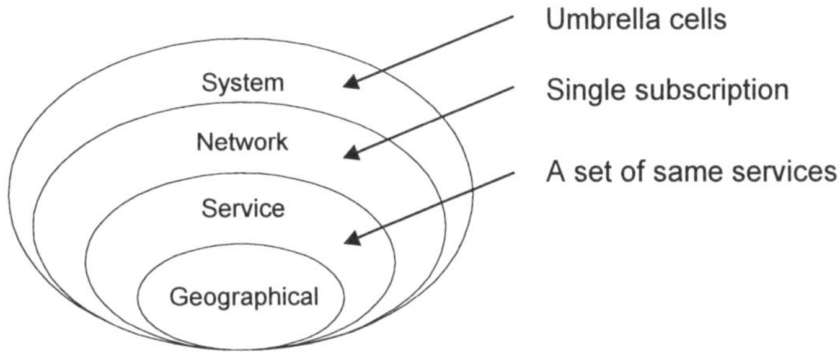

Figure 3. The levels of integration as defined in (ETR 093).

Geographical integration

In geographical integration the satellite and terrestrial system are based on totally different techniques and do not necessarily provide compatible services (ETR 093). Moreover, the subscribers using the systems perceive them in a different manner. For both systems an own subscription is needed and the caller has to know, in which system the user is currently reachable as there are no means for routing in between the systems.

Service integration

In service integration the system parameters of the satellite system are such that they can support a common set of services (ETR 093), which allows, e.g., end-to-end voice connection. Clearly, there can be limitations in terms of services offered and in their quality at satellite system when compared to the terrestrial system.

Figure 4 illustrates a call scenario for service integration. The called person, Mikko, has subscriptions and terminals for both satellite and terrestrial system. The first call is directed to Mikko's terrestrial mobile subscriber number (MSN1 in *Figure* 4). The call fails because Mikko is served by the satellite and there is no routing between the systems. Therefore, the calling person, Hans, has to place a second call, but this time towards Mikko's satellite subscriber number (MSN2 in *Figure* 4). Eventually, the second call is connected and the two persons have an end-to-end connection for service C, because the two systems have that service in common. However, service D in the satellite system is available in a limited way only. For example, the highest bit rate for the data services can be lower than in the terrestrial system.

Network integration

The network integration can be characterised by a single subscription and that the caller does not need to know in what system the user is currently located (ETR 093). The systems will automatically take care of the necessary routing between the systems. This is similar to the mobility between GSM networks. At the network side, the existing terrestrial switching equipment and databases can be reused by the satellite system or shared between the systems, which both allow savings in design costs.

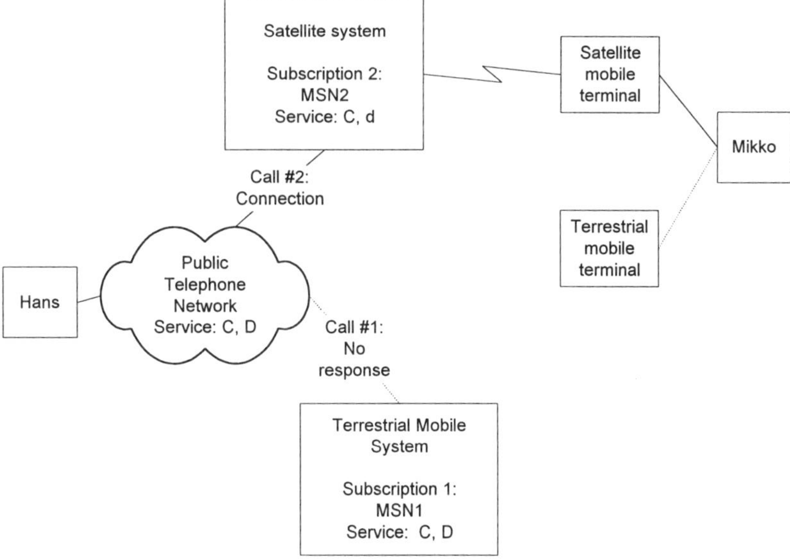

Figure 4. A call scenario for service integration.

System integration

At the system level the satellite cells are considered as umbrella cells to the terrestrial system (ETR 093). The satellite access is thus an integral part of the terrestrial network. From the point of view of the terrestrial system, the satellite cells are just like any terrestrial cells and handovers between all cells must be provided in a seamless fashion. Basically, at this level all the functionality and the protocols could be the same in both components, except low layer radio resource control. This may be needed, for example, due to different frequency bands of operation.

3.2 Terminal aspects

An important aspect of integration considers the terminal. A basic requirement is that the two systems could be accessed with a single piece of an equipment. In general, that sort of an equipment can be called dual mode terminal discussed, e.g., in (Kuisma, 1997). As a minimum, the integration may represent "placing two terminals inside a single cover". This would mean that the two access modes could possibly share, e.g., keyboard, microphone, speaker, display and battery. All other parts of the terminal would be duplicated and they would be based on different technologies. Ideally, the terminal would have most functionality and physical parts in common.

4. INTEGRATION OF SATELLITES INTO GSM

4.1 Situation at the beginning of GSM standardisation

The mobile satellite aspects were not included in the initial GSM standardisation. When GSM standardisation was started only geostationary satellite systems with large (e.g. suitcase size) terminals were available or planned. The GMPCS systems were introduced only after GSM was in a mature state. Of the emerging systems, so called S-PCNs (satellite personal communication network) may eventually have relevance to GSM, as they represent circuit switched voice and their service offering is aimed to be comparable to GSM.

As a result of the emerging S-PCNs, a placeholder for possible future work with satellite specific aspects was included into the GSM standardisation as a phase 2+ work item called "Interworking with mobile satellite systems". The work item is described in (GSM 10.00). However, the introduction of the work item does not mean that any satellite aspects were inevitably included in the GSM standard.

4.2 Service integration with GSM

With respect to service integration, the need for two subscriptions is here regarded as a drawback. This is because GSM supports full-fledged international roaming between GSM networks. However, appropriate conditional call forwarding could be utilised to achieve "automatic" routing in between the systems. From the GSM point of view, service level integration appears as interworking toward public networks, which will provide interconnection for the systems.

The interconnection arrangement for service integration is illustrated in *Figure 5*. For example, the protocols such as mobility management (MM), connection management (CM), basestation subsystem management part (BSSMAP), and radio resource management (RR) would be proprietary to the satellite system. Also, the satellite system would not implement GSM mobile application part (MAP) protocol, which supports seamless mobility (or roaming) between GSM networks. The corresponding GSM protocols are described generally, e.g., in (Mouly et al, 1992).

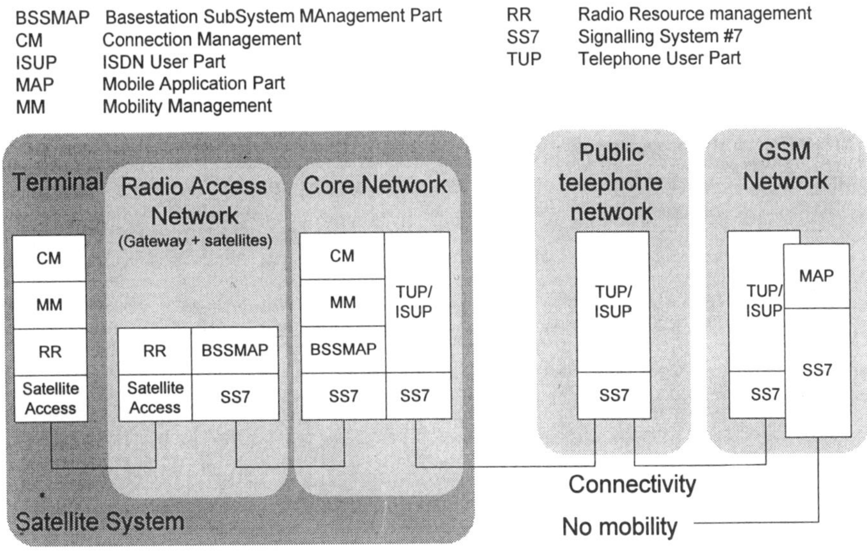

Figure 5. Scenario for service level integration.

4.3 Network integration with GSM

With respect to the network integration, the adoption of the GSM MAP protocol is required to support seamless roaming between the two systems having a single subscription. This is illustrated in *Figure 6* in terms of protocols. Adoption of the GSM MAP is possible because it is independent of the radio access (Verkama et al, 1996). In practice, the satellite system should either implement fully the GSM MAP protocol or it should provide GSM MAP interworking towards GSM with some other comparable network protocol.

The adoption of the MAP protocol means that the satellite system can reuse or share the GSM switching infrastructure and registers at network level. Here, reusing means that the core network (in particular GSM MSC/VLR) is devoted to the satellite system only, i.e., a separate network with its own infrastructure is formed. Therefore, deviations from GSM in the interface between the radio access network and core network (called A-interface in GSM) are possible. However, when *Figure 6* is considered the A-interface has to comply to GSM at least partly when MM and CM protocols are considered. For example, the authentication method in the satellite system has to be compliant to GSM.

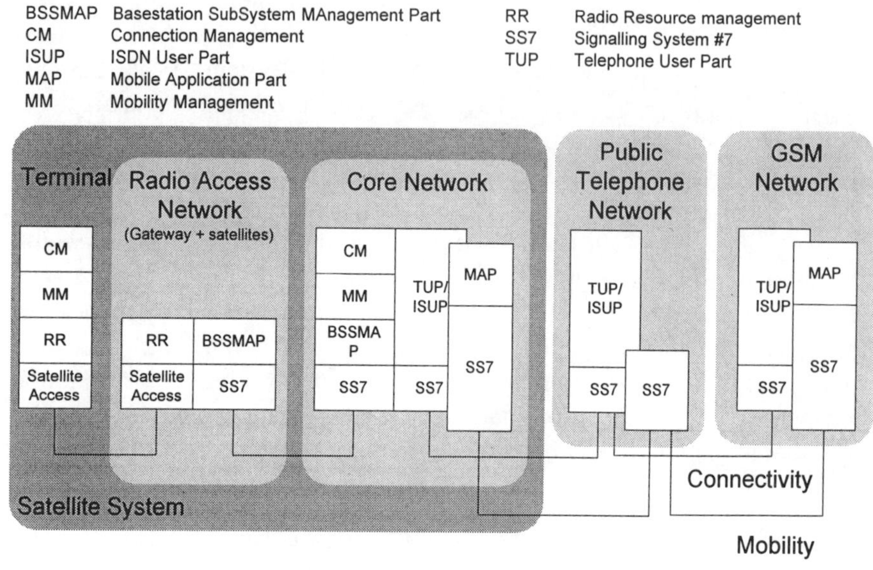

Figure 6. Scenario for network level integration.

In a shared configuration, a single GSM MSC/VLR provides connections both into a GSM basestation subsystem (BSS) and a satellite gateway (see *Figure 7*). This requires that the gateway has to comply fully to the GSM A-interface standard to avoid changes in the GSM equipment. However, the radio access network and its internal protocols and procedures, especially for satellite access and RR, would still be proprietary to the satellite.

The capability of roaming with single subscription also requires the adoption of the GSM subscriber identity module (SIM) and GSM SIM-ME (mobile equipment) interface in the satellite system. Ideally, a GSM SIM could be used to access the satellite. In practice, however, the SIM may have to contain a satellite specific application, which requires replacing of the existing GSM SIM with a dual mode SIM. The roaming between the systems could thus be based on a single SIM card or on dual mode terminal (with SIM) capable of operating in both domains.

The location management is an important area especially for those satellite systems that have adopted GSM based NSS. One particular problem related to non-geostationary systems utilising a GSM NSS is the deployment of the GSM location areas that are comprised of sets of fixed cells. In non-geostationary systems, however, the nature of the radio network is quite different from the GSM due to the satellite motion relative to the surface of the Earth. As a result, it is possible that the GSM location area solution as such is not feasible for the satellite system.

One obvious solution is to locate terminals by their geographical co-ordinates rather than by using GSM location areas. For example, to avoid storage of information specific to the satellite in the GSM VLRs, such as geographical co-ordinates, a satellite location register (SLR) is proposed by Argenti et al (1996). If SLR was not introduced, changes would possibly be needed, e.g., in the GSM VLR. *Figure 7* illustrates the registers as part of the system. The SLR would contain temporary non-GSM information required by the satellite component. Moreover, it would implement interworking functionality by performing transformation between geographical co-ordinates and GSM location area for location update and paging.

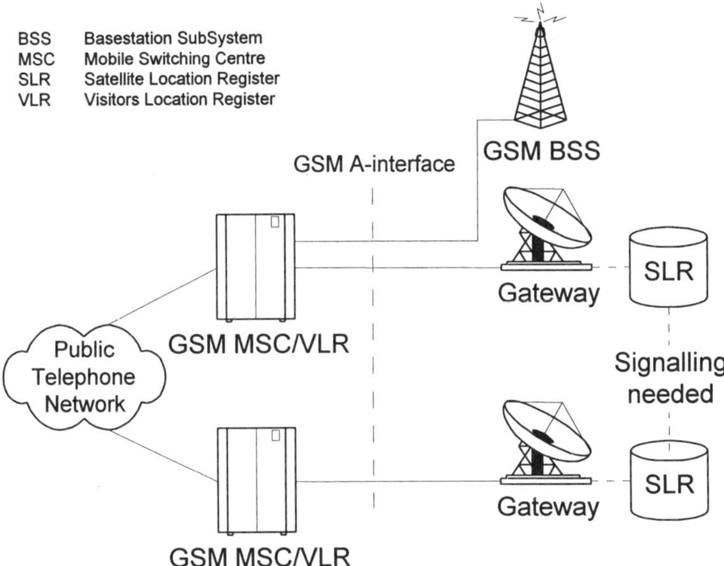

Figure 7. An arrangement with a register in a gateway.

Even if changes of the GSM NSS were avoided, the SLR would add the complexity to the gateway due to increased functionality. Moreover, a direct communication between gateways connected to different MSCs would be needed, e.g., to facilitate deletion of subscriber data from the old gateway at the registration into a new gateway (see *Figure 7* and (Argenti et al, 1996, p. 757)). This functionality is similar to the one supported by the GSM MAP protocol between two VLRs and the HLR.

4.4 System integration with GSM

System integration with GSM is here defined as interaction between GSM and satellite when the terminal is in connected mode. Handover during a call between GSM and satellite access is a good example on such an interaction. The motivation of introducing such handovers is to allow connection continuity, e.g., when a dual mode GSM/S-PCN terminal leaves the GSM coverage. These handovers are discussed, for example, in (Del Re et al, 1995) and (Zhao et al, 1996).

An inter-system handover from satellite to GSM could basically appear to GSM as an external or inter-MSC handover, depending on the satellite system architecture. When *Figure 6* is considered, the gateway or satellite system would have to adopt at least part of the GSM BSSMAP protocol to support the handovers. Also, the satellite system should implement functionality to initiate the handover toward the target cell in the desired GSM network.

From the GSM point of view, handover from GSM to a satellite system seems to require changes in the GSM infrastructure. If the general GSM procedures for handover are followed, the nature of the requirements on GSM resemble quite much those faced with "multiband operation" (GSM 03.26), which represents system integration with GSM 900 and GSM 1800. Similar to multiband operation, the terminal should receive a list of neighbouring cells belonging to the other system to avoid random search of the satellite cells. Also, the terminal should inform the network about its capabilities to operate in the other band or system. These would have to be standardised for GSM.

In the case of a non-geostationary system, it is possible that the satellite neighbouring frequencies are constantly changing, which would make the interoperability more demanding. The GSM BSS would implement functionality to trigger the handover based on the measurements performed and reported by the terminal. The request for handover targeted to satellite component should include the location of the terminal in some applicable form for the satellite system. For example, the BTS location or the actual location of the terminal could be utilised (Zhao et al, 1996) especially in non-geostationary systems.

5. EMERGING S-PCN SYSTEMS WITH GSM-BASED NSS

There are several emerging S-PCN systems that aim to offer services that would match closely to GSM. To achieve this, the systems have to be comparable to the GSM on issues such as QoS, speech quality, delay, terminal properties, and user charges. Moreover, access into both systems with single subscription is an important requirement. Ideally, the S-PCNs would offer outdoor service extension for terrestrial personal communications systems such as GSM. In the following, S-PCNs that have adopted GSM platforms or protocols are discussed.

Globalstar (Hirshfield, 1996), ICO (Ghedia, 1995), Iridium (Brunt, 1996) and Odyssey (Baird et al, 1996) (in alphabetical order) are global non-geostationary S-PCN proposals that appear to be often referenced. According to Huber (1995) Globalstar, ICO and Iridium will have GSM based platforms in their systems. The adopted platform consists of the GSM NSS, that represents core network in the system (*Figure 1*). According to (Odyssey), also Odyssey has adopted GSM type of architecture. There are also regional S-PCN systems being planned for Asia, Africa and the Middle East. These systems include at least ACeS, Africom, Agrani, AMSC (II), APMT, EAST and Satphone (Evans, 1997). Certain of these systems may plan to adopt GSM platforms and protocols. However, the geostationary (like possibly some non-geostationary) systems are not very well know in the technical literature.

The non-geostationary S-PCN systems with the GSM NSS represent network level integration with GSM. The interconnection arrangements of ICO, Iridium and Odyssey appear to be roughly such that they all form a network connecting either their satellites, gateways, or core network entities. Globalstar, however, aims at connecting both a gateway and the GSM BSS into the same GSM MSC (see (Mangir, 1997, p. 36) and *Figure 7*). Seemingly, such a gateway has to comply fully to the GSM A-interface. Iridium implements direct communication links between the satellites. As a result, the space segment forms a network for the system, which in principle allows operation with only one gateway in the system. In other systems the number of gateways has to be at least comparable to the number of satellites to allow operation in the targeted service area.

The emerging S-PCN systems aim at introducing handheld dual mode GSM/S-PCN terminals. They are also expected to offer dual mode terminals for other terrestrial standards. The dual mode terminals are important for the systems, when service extension for the terrestrial systems and addressing of the existing GSM subscribers are considered.

The potential user groups or applications addressed by these emerging global systems include roughly international business persons, national roamers and national fixed service. The business persons would utilise these services while travelling in underdeveloped areas or in regions with incompatible terrestrial standards. This is the most robust group of potential users as the cost is typically not an issue for these people. The national roamers would consists, for example, of people using the services while staying at their summer cottage outside the terrestrial coverage. The national fixed rural service would be utilised, for example, by people living in distant rural areas that are currently not or underserved by communications systems. For these users, the systems could provide an extension for any fixed service like voice telephony. However, at least part of these users may have low income. As a result, their ability to pay for the service may be limited. For this kind of fixed applications, regional geostationary systems are particularly suitable, as geostationary satellites remain fixed to the surface of the Earth

5.1 GSM NSS considerations

In GSM the radio aspects are well isolated into the BSS part of the system. Therefore, competencies required for GSM based S-PCN NSS implementation appear quite similar to those needed in GSM on a general level. Obviously, proprietary solutions, e.g., for the GSM A-interface may be required in the actual implementation. In the following, aspects that appear specific to the S-PCN NSS are discussed briefly.

The capacity requirements of a particular S-PCN system can be quite different from those encountered in a conventional GSM NSS. These requirements are projected into the GSM NSS in terms of switching and the number of users served by the NSS. Parameters affecting the requirements on the NSS capacity include, for example, the total number of satellites in the system, the capacity of an individual satellite, the system architecture, and anticipated user numbers and traffic density. For example, in a MEO system with around 10 gateways it is well possible that the GSM NSS has to be capable of serving the subscribers on an area comparable to the size of the Europe.

The radio-interface limitations may require introduction of new S-PCN specific bearer services that may have effect on the interworking function of the GSM NSS. Also new, non-GSM services may be introduced. For example, a service that could offer, e.g., paging like indoor connectivity, would provide added value for the systems. The paging could be utilised, for example, to alert the user of an incoming call that can only be completed, provided the user moved into a more favourable environment in terms of line-of-sight to the satellite. Also, the requirements on S-PCN legal interception can be quite different from those faced in GSM. For example it may well be possible that the authorities in a particular country want to intercept their citizens', but the gateway is in another country. This is of course a regulatory problem as such, but it may also require new solutions in the GSM NSS.

6. DISCUSSION

The first geostationary and non-geostationary S-PCN systems are planning to start their operation during 1998 (Nourouzi et al, 1996). Prior to that, the systems have many technical problems to solve, such as the design of the handheld terminal antenna. Issues affecting the business success of the systems include at least securing the initial financing, time to market, and the cost of using the system, which includes also terminal cost.

An interesting matter with the S-PCN systems aiming to complement terrestrial systems is whether end users with dual mode terminals will expect the same QoS and services as in terrestrial mobile systems. An alternative is that S-PCN service is perceived as something different and possibly an overall degradation is tolerated. For example, S-PCNs cannot provide indoor coverage in general. It is possible that some users are not very eager to go, e.g., outdoors to complete a call after receiving a paging message for an incoming call. The same applies, e.g., for downloading times in data services. According to Abrishamkar et al (1996) the maximum rates available in S-PCNs (Globalstar, ICO, Iridium and Odyssey) range from 2.4 to 9.6 kbit/s. Naturally, S-PCNs may also introduce new enhanced data services for their users in the future.

One further aspect relevant for the S-PCNs, is the development in the terrestrial mobile systems in terms of coverage. The extent of the coverage of the terrestrial services has surpassed expectations that were made in the early nineties. Also, multi mode or multi band operation between different terrestrial systems may offer even larger footprint with single subscription. For example a "GSM world phone" operating at GSM 900/GSM 1800/GSM 1900 frequency bands would offer service in Europe, Asia and the USA. This may influence, e.g., on the number of international business users that would use an S-PCN system due to incompatible standards in different regions.

REFERENCES

Abrishamkar F. and Siveski Z., (1996). "PCS global mobile satellites", IEEE Communications magazine, September, 1996, pp. 132-136.

Ananasso F., (1996). "Sinergy between satellite, wireless and wireline technologies in the global information infrastructure", in 7th international network planning symposium "planning networks & services for the information age", vol. 2, pp. 49-54, Sydney, November 24-29, 1996.

Argenti F., Cappelletti L., Del Re E. and Ferro A., (1996). "Integration of satellites into GSM: Signalling flow analysis", in Proceedings of the IEEE International Conference on Universal Personal Communications, vol. 2, pp. 755-759, Cambridge, MA, September 29 - October 2, 1996.

Baird T. and Bush W., (1996). "Odyssey system overview", Professional Program Proceedings. ELECTRO `96, pp. 15-25, Somerset, New Jersey, April 30-May 2, 1996.

Brugel E. W., (1994). "On-board processing and future satellite communication services", Space Communications, vol. 12, iss. 3-4, 1994, pp. 121-174.

Brunt P., (1996). "Iridium - overview and status", Space communications, vol. 14, iss. 2, 1996, pp. 61-68.

Del Re E. and Iannucci P., (1995). "The GSM procedures in an integrated cellular/satellite system", IEEE Journal on selected areas in communications, vol. 13, no. 2, 1995, pp. 421-430.

ETR 093. Possible European standardisation of certain aspects of satellite personal communication networks (S-PCN) phase 1 report. ETSI, technical report (ETR 093), September 1993.

ETR 12.02. Technical characteristics, capabilities and limitations of mobile satellite systems applicable to the Universal Mobile Telecommunications System (UMTS). ETSI, draft technical report UMTS 12.02 (DTR/SMG-051202), Version 2.0.2, September 1994.

Evans J. V., (1997). "Satellite systems for personal communications". IEEE Antennas and Propagation Magazine, vol. 39, no. 3, June 1997, pp. 7-20.

Gaffney L. M., Hulkower N. D. and Klein L., (1996). " Non-GEO mobile satellite systems: a risks assessment", Space Communications, vol. 14, iss. 2, 1996, pp. 123-129.

GSM 03.26. Multiband operation of GSM/DCS 1 800 by a single operator. ETSI, technical report GSM 03.26 (ETR 366), Version 5.1.0, August 1997.

GSM 10.00. Feature descriptions. ETSI, technical report GSM 10.00 (SMG-TR 010), Version 5.2.0, July 1997.

Hirshfield E., (1996). "The Globalstar system: breakthroughs in efficiency in microwave and signal processing technology", Space communications, vol. 14, iss. 2, 1996, pp. 69-82.

Huber J. F., (1995). "Mobile goes global", Telcom report international, vol. 18, iss. 3, 1995, pp. 7-10.

Kuisma E., (1997). "Technology options for multi-mode terminals", Mobile communications international, iss. 38, February 1997, pp. 71-74.

Mangir T. E., (1997). "Wireless via satellite: Systems for personal/mobile communication and computation", Applied microwave & Wireless, vol. 9, iss. 1, 1997, pp. 24, 26-8, 30, 32-4, 36, 38, 40, 42.

Mouly M. and Pautet B., (1992). "The GSM System for Mobile Communications", Published by the authors, ISBN 2-9507190-0-7, 1992.

Odyssey. Odyssey technical sheet. TRW Odyssey Telecommunications International, Inc.

Nourouzi A. and May A., (1996). "LEOs, MEOs and GEOs: the market opportunity for mobile satellite operators", Ovum Ltd., London, December 1996.

Pritchard W. L., Suyderhoud H. G., and Nelson R. A., (1993). "Satellite communication systems engineering", 2nd ed. Englewood Cliffs, NJ: Prentice-Hall, 1993.

Verkama, M., Söderbacka, L., and Laatu, J. (1996). "Mobility management in the third generation mobile network," in Proceedings of IEEE Global Telecommunications Conference, Vol. III, pp. 2058-2062, London, November 18-22, 1996.

Zhao W., Tafazolli R. and Evans B. G., (1996). "Positioning assisted inter-segment handover for integrated GSM-satellite mobile system", in Satellite systems for mobile communications and navigation (Conf. Publ. No. 424), pp. 124-128, London, May 13-15, 1996.

Chapter 4

DECT/GSM Integration

Enhancement of GSM using DECT

JÖRG KRAMER, ARMIN TOEPFER
Mannesmann Mobilfunk GmbH

Key words: DECT, GSM.

Abstract: In the last years the Global System for Mobile communication, GSM, has gained a tremendous success around the world. Meanwhile more than 250 operators offer their GSM services to over 66 Mio customers in 110 countries. Before this success could be imagined, work was started in ETSI to specify first drafts for the interworking of Digital Enhanced Cordless Telecommunications, DECT, to GSM. The extension of GSM using DECT as a complementary air interface was seen as a step towards more capacity and enhanced radio performance, using synergies and the best characteristics of both systems. During the last years different access standards for DECT were elaborated, including the DECT/GSM Interworking Profile, GIP. This profile allows the attachment of DECT to GSM, while using all specified GSM phase 2 services, like supplementary services, facsimile, bearers services and the short message service (SMS) via the DECT air-interface. The following chapters describe the access technology DECT, some of the specified access profiles, different technical DECT/GSM integration scenarios and indicates the advantages and drawbacks of the usage of DECT as an extension to GSM in the residential, business and public environment.

1. DECT - OVERVIEW

In the year 1988 the European Telecommunication Standardization Institute (ETSI) started to develop a new digital cordless standard called DECT. The basic standard is available since 1992. The DECT frequency band is reserved for DECT applications in all 43 CEPT countries. Since

1996 DECT stands for **D**igital **E**nhanced **C**ordless **T**elecommunications and gains global availability. DECT is a powerful and versatile digital radio access technology, that makes efficient use of radio spectrum and supports applications such as access to residential, business, public and local loop environments. Meanwhile DECT is a success not only in Europe, it is also used in Argentina, Australia, Austria, Bahamas, Bahrain, Bolivia, Brazil, Cambodia, Chile, China, Columbia, Cuba, Hong Kong, Hungary, India, Indonesia, Israel, Mexico, New Zealand, Norway, Romania, Singapore, South Africa, Spain, Sri Lanka, Thailand, the United Arab Emirates, Uruguay, Venezuela and some more. (Trials and commercial systems)

In the U.S., Canada and Mexico DECT is re-used as Personal Wireless Telecommunications, PWT, incorporating just some modifications on the physical layer.

1.1 Characteristics and Structure

1.1.1 DECT Characteristics

The key characteristics of DECT are:

- DCS/DCA (Dynamic Channel Selection / Allocation), seamless handover, variable bandwidth
- High traffic capacity - up to 10 000 Erlang per km² per floor
- Adaptive Differential Pulse Code Modulation - ADPCM, for a very high speech quality
- Data service from 24 kbit/s up to 552 kbit/s (2Mbit/s in preparation)
- pico-cellular coverage with small 50 metre cells and a tested outdoor radius of up to 3 - 5 km
- Service transparent access for a variety of environments like:

- Residential (Connected to the PSTN/ISDN)
- Business (Wireless multi-cell Private Telecommunication Networks, PTNs)
- Public Access (for direct access to a public network)
- Wireless Local Loop (e.g. replacement of wired local loop)
- Wireless Local Area Networks (LAN)
- Cellular extension (GSM)

Recently the basic DECT standard was expanded to cover those applications described.

1.1.2 DECT Structure

DECT is a pico-cellular system that operates with 10 carriers in the assigned 20 MHz frequency band between 1880 and 1900 MHz (Frequency band used in Europe, Extension band 1900-1920 MHz possible). It is a Time Division Multiple Access (TDMA), Time Division Duplex (TDD) system. One frame contains 24 normal slots providing a total datarate of 1152 kbit/s. The first 12 slots are for the downlink (basestation -> mobile station) and the second 12 slots are for the uplink (mobile station -> base station). The basic channel structure can be seen in figure 1.

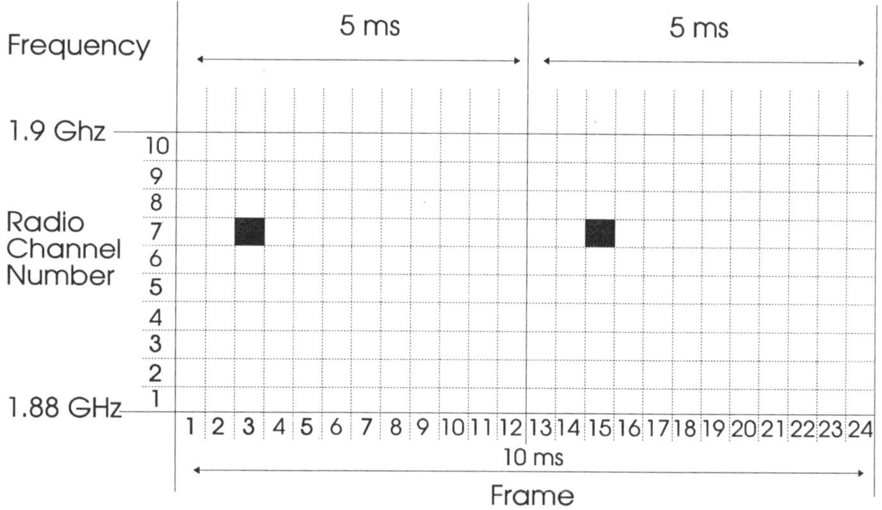

Figure 1: DECT Frame Structure

In total 120 duplex channels for the protected and unprotected mode are available for data transmission. The channel rate for DECT data in the protected mode is 24 kbit/s per slot supporting a very low bit error rate (BER: $<10^{-10}$), for ADPCM speech transmission the 32 kbit/s unprotected mode is used as shown in figure 2.

With the dynamic allocation of channels, advanced connections are possible, e.g. the combination of two slots providing 64 kbit/s for transparent support of ISDN connections, and even the combination of up to 23 slots (uni-directional: 552 kbit/s), as specified in the DECT data profiles (see also chapter 1.2.3 on page 290). Figure 2 shows the DECT slot structure embedded in the frame and the multiframe.

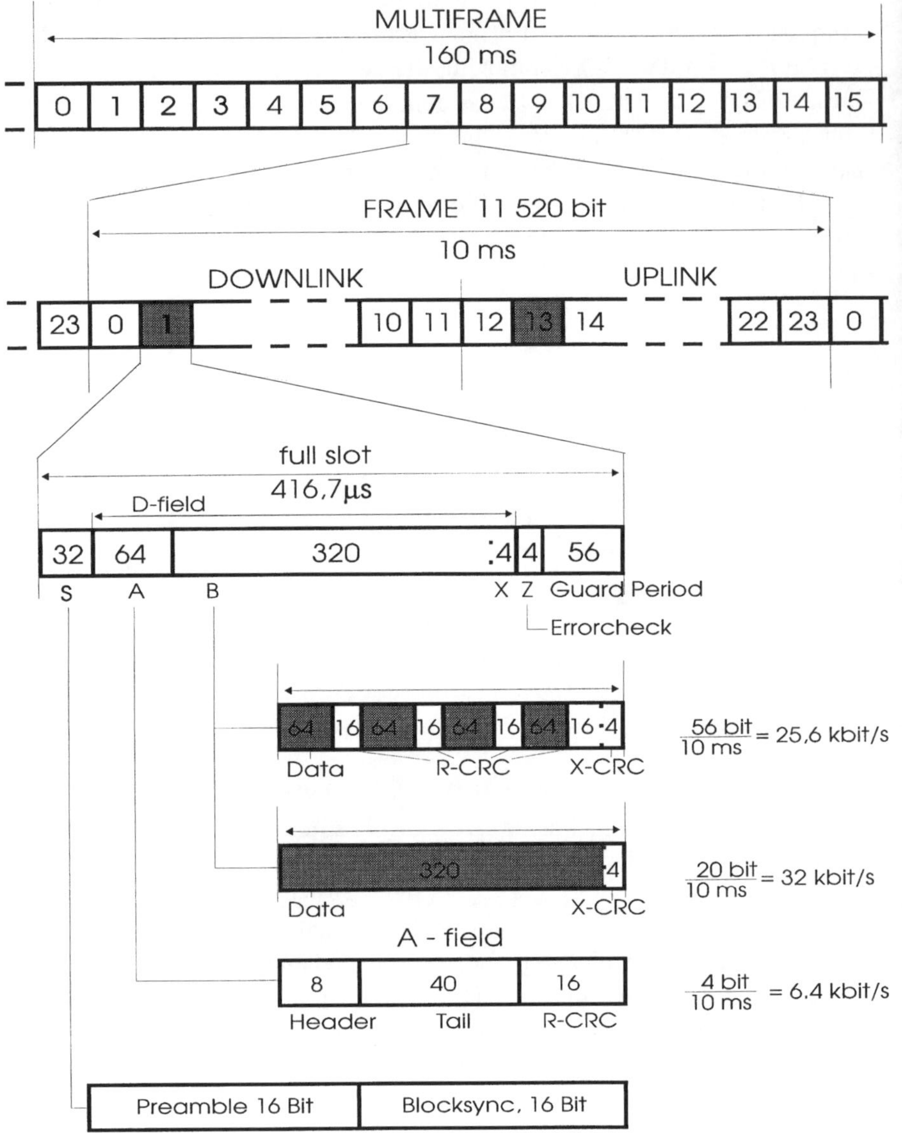

Figure 2: DECT Slot Structure

1.2 Standards and Access Profiles

1.2.1 DECT Basic Standard

The relevant DECT basic standard is the ETS 300 175, Part 1 to 9. These nine parts are shortly described in the next clauses:

1.2.1.1 Part 1: Overview 175-1

The overview contains an introduction to the complete ETS. It includes a description of the system and the protocol architecture and a vocabulary of terms.

1.2.1.2 Part 2: Physical layer 175-2

The PHysical Layer (PHL layer) specifies radio parameters such as the frequency, timing and power values, the bit and slot synchronisation and the transmitter and receiver performance.

1.2.1.3 Part 3: Medium access control layer 175-3

The Medium Access Control layer (MAC layer) specifies three groups of MAC services. These are the broadcast message control service, the connectionless message control service and the multi-bearer control service. It also specifies the logical channels, which are used by the above mentioned services, and how they are multiplexed and mapped on to the physical channels.

1.2.1.4 Part 4: Data link control layer 175-4

The Data Link Control layer (DLC layer) specifies two groups of DLC services. These are the services for the control plane and the services for the user plane.

For the control plane a point-to-point service and a broadcast service are defined. The point-to-point service can operate in acknowledged, or unacknowledged, mode and provides addressing, frame delimiting, error control, flow control, segmentation of network layer information fields, fragmentation of DLC frames and connection handover.

For the user plane the transparent unprotected service, the frame relay service, the frame switching service and the rate adaption service are defined.

1.2.1.5 Part 5: Network layer 175-5

The NetWorK layer (NWK layer) specifies the functions for the link control, the call control, the supplementary services, the connection oriented message service, the connectionless message service and the mobility management. For the above mentioned groups it contains the procedures, messages and information elements.

1.2.1.6 Part 6: Identities and addressing 175-6

The identities and addressing specify the main identities and addresses which are used in DECT. They are divided into the following four categories: fixed part identities, portable part identities, connection related identities and equipment related identities. Several of the fixed part identities and portable part identities are allocated centrally in order to maintain global uniqueness for these identities.

1.2.1.7 Part 7: Security features 175-7

The security features standard specifies the overall security architecture for DECT, the types of cryptographic algorithms required, the way in which they are to be used and the requirements for integrating the security features provided by the architecture into the DECT air interface. It also describes how the features may be managed and how they are related to certain DECT fixed systems and local network configurations.

1.2.1.8 Part 8: Speech coding and transmission 175-8

The speech coding and transmission specifies the requirements for DECT equipment which includes all the necessary functions to provide real-time two-way speech conversation. It defines the speech encoding algorithm and the detailed speech performance characteristics such as sensitivity, frequency response, sidetone, terminal coupling loss, distortion, variation of gain with input level, out of band signals, noise, acoustic shock, delay and network echo control.

A 3,1 kHz telephony teleservice conveyed over a DECT link which is capable of being connected (directly or indirectly) to the public network access point shall comply with the requirements in Part 8.

1.2.1.9 Part 9: Public access profile (PAP) 175-9

The public access profile specifies the mandatory (minimum) and optional requirements for DECT equipment offering, or using, public access services. This profile specifies an interoperable subset of the protocol independently for fixed parts and portable parts, by detailed references to the other parts of the ETS.

Practically the PAP is superseded by the new developed Generic Access Profile, GAP, which offers interoperability for residential, business and public access.

1.2.2 DECT Profiles

DECT is an access technology to networks like ISDN, PSTN, PTN, GSM or X.25. The DECT base standard ETS 300 175 only describes capabilities, but does not specify exactly which features or procedures are mandatory for each application. Therefore, in order to provide for interworking between DECT terminals supporting the same application, specific access and interworking profiles have been developed or are still under elaboration.

- PAP - Public Access Profile (see chapter 1.2.1.9)
- GAP - Generic Access Profile (see chapter 1.2.2.1)
- DECT Data Profiles (see chapter 1.2.3)
- RAP - Radio in the local loop Access Profile
- CAP - CTM Access Profile
- IAP/IIP - DECT/ISDN Interworking Profile
- GIP - DECT/GSM Interworking Profile (see chapter 1.2.4)
- ISDN-based-interface DECT/GSM Interworking Profile (see chapter 2.5.2)

For the interworking of DECT and GSM, four profiles are described in the following: The Generic Access Profile (GAP), the DECT Data Profiles, the DECT/GSM Interworking Profile (GIP) and the ISDN-based DECT/GSM Interworking Profile.

1.2.2.1 GAP - Generic Access Profile

The **GAP** (ETS 300 444) defines general interoperability requirements for any private or public DECT application supporting a 3,1 kHz telephony teleservice. The GAP enables the linking of residential, business and public access environments, offering the ability to use a single terminal in different locations and with different networks. It includes mobility management and security features.

As those access environments may be different, specific requirements are defined for the respective fixed part attached to the network.

The GAP is adopted by ETSI and the appropriate Technical Basis for Regulation, TBR 22, is valid since end of 1997.

1.2.3 DECT-DATA Profiles

The DECT Data Profiles make use of the 'toolbox' characteristic of the DECT base standard. A family of data profiles complete the open standard character of DECT, by ensuring inter-operability between products from different manufacturers. They all exploit the powerful lower-layer data services of DECT, which are specifically oriented towards LAN, multimedia and serial data capabilities, but each member of the profile family has been optimised for a different kind of user service. The different profiles are created in modules to enable an economical and efficient implementation.

The family of class 2 profiles are described briefly in the next clauses. Class 2 profiles are designed to support mobility and roaming applications, both in the public and private environment. Class 1 Profiles are available, but don't support mobility and are therefore not described in this text.

- ETS 300 701: Data Services Profile, Generic Frame Relay Service with Mobility (**Service Types A and B, Class 2**) supports low-speed frame relay with data rates up to 24,6 kbit/s (Type A), and high-performance frame relay, throughput up to 552 kbit/s (Type B).

- EN 300 651: Data Services Profile, Generic data link service (**Service Type C, Class 2**) specifies a Link Access Protocol (LAP) service and extends the Data Stream service into environments, such as public services, where significant mobility is a characteristic;

- EN 301 238: Data Services Profile, Isochronous data bearer services with roaming mobility (**Service Type D, Class 2**) allows the support of transparent and isochronous connections of 32 kbit/s and an rate adaptation service with up to 28,8 kbit/s.

- ETS 300 757: Data Services Profile, Low Rate Messaging Service (**Service type E, Class 2**) specifies a Link Access Protocol (LAP) service suitable for non-transparent transfer of asynchronous character streams and is intended for usage in private and public roaming environments. It shall be used for the support of SMS and similar services.

- ETS 300 755: Data Services Profile, Multimedia Messaging Service with specific provision for Facsimile services (**Service type F, Class 2**) creates high level inter-operability for a range of Telematic services, including fax, through a multi-media file transfer mechanism built on the data stream service, with full support for roaming and public services.

1.2.4 GIP - GSM Interworking Profile

The GIP (DECT - **G**SM **I**nterworking **P**rofile) is a set of seven standards that describe how DECT shall be connected to a GSM PLMN via the A-Interface.

1. Access and mapping (protocol/procedure description for 3,1 kHz speech service), (ETS 300 370);
2. General description of service requirements, functional capabilities and information flows, (ETS 300 466);
3. GSM-MSC/DECT-FP Fixed Interconnection, (ETS 300 499);
4. GSM Phase 2 Supplementary Services implementation, (ETS 300 703);
5. Short message services, point to point and cell broadcast, (ETS 300 764);
6. Implementation of bearer services, (ETS 300 756);
7. Implementation of facsimile group 3, (ETS 300 792).

The concept of DECT/GSM interworking is basically that of accessing a GSM network using a DECT air interface and offering all GSM phase 2 services, where the DECT Fixed Part/InterWorking Unit (FP/IWU) emulates the GSM BSS. No changes on the GSM side are necessary. The interworking profiles use the existing DECT air-interface structure with additional requirements like the GAP and the DECT-Data profiles E, C and F.

The list of the relevant standards and their appropriate status can be found in chapter 4.

An overview on assignable GIP-applications and GIP-scenarios is given in chapter 2.2 and 2.5. A closer look on the DECT/GSM dual-mode terminal is given in the chapter 2.5.4.

The mapping of GSM A-interface messages and the appropriate DECT messages is ensured by the IWU that can be seen in the next figure.

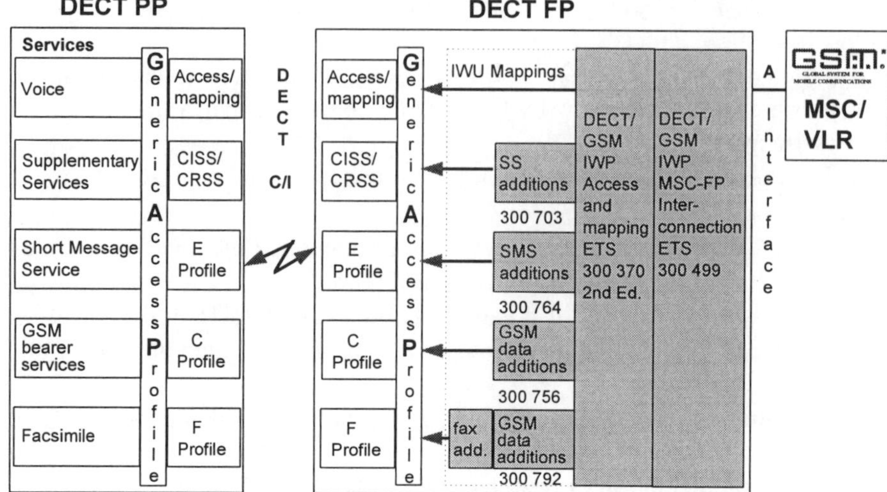

Figure 3: GSM Interworking Profile - GIP

ETS 300 370 describes main functions like GSM authentication, derivation of the DECT ciphering key from the respective GSM chipher key, usage of GSM-IMSI and -TMSI, GSM Location Area Identity (LAI) and the support of subscription management by the use of GSM SIM.

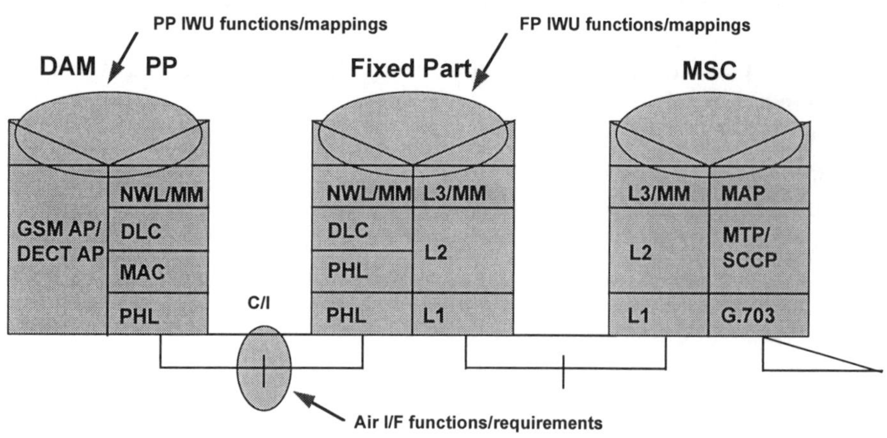

Figure 4: Interworking Units (IWU) for the GIP

As we see two IWUs are required, one between the MSC and the DECT Fixed Part (FP) and one in the DECT portable part (PP). The first IWU provides the mapping between a subset of GSM layer 3 to the corresponding DECT NWK layer protocols. The PP IWU provides mapping between a subset of the DECT network layer protocol and the GSM application (SIM).

ETS 300 466 covers the requirements for the provision of GSM services over the DECT air-interface. It includes a general description of service requirements, functional capabilities and information flows for teleservices, bearer services and supplementary services.

ETS 300 499 covers the interconnection of the GSM MSC with the DECT fixed part (FP) that is based on the A-interface.

ETS 300 703 describes the support of all GSM phase 2 supplementary services.

The following GSM phase 2 supplementary services are supported by the GIP:

Table 1. GIP Supplementary Services

Call Offering SS	**[GSM 02.82]**	
CFU		Call Forwarding Unconditional
CFB		Call Forwarding on Mobile Subscriber Busy
CFNRy		Call Forwarding on no Reply
CFNRc		Call Forwarding on Mobile Subscriber Not Reachable
Call Completion SS	**[GSM 02.83]**	
CW		Call Waiting
HOLD		Call hold
Multi Party SS	**[GSM 02.84]**	
MPTY		Multi Party
Community of Interest SS	**[GSM 02.85]**	
CUG		Closed User Group
Charging SS	**[GSM 02.86]**	
AoCI		Advice of Charge Information
AoCC		Advice of Charge Charging
Call Restriction SS	**[GSM 02.88]**	
BAOC		Barring of All Outgoing Calls
BOIC		Barring of Outgoing International Calls
BOIC-exHC		Barring of Outgoing International Calls except those directed to the Home PLMN
BAIC		Barring of All Incoming Calls
BIC-Roam		Barring of Incoming Calls when Roaming outside the home PLMN

The general approach is based on the functional protocol that shall be implemented in the DECT-GIP terminal.

For the support of GSM phase 2 supplementary services generally the DECT {FACILITY} message together with <<IWU-TO-IWU>> information elements containing GSM facility elements are used for different Connection Independent Supplementary Services (CISS) and Connection Related SS (CRSS) procedures. With this approach a transparent support of services is ensured and interworking to new services, like GSM phase 2+ supplementary services can be introduced very simply.

For CISS, the DECT Call Control (CC)-connections with the Basic service information element and call class "supplementary service call set-up" is used for being able to support all services also during handover between different DECT clusters (external handover).

Hold and Retrieve procedures, enabling supplementary services like Call Hold and Multi Party, use the separate message approach mapping the messages {HOLD}, {HOLD-ACK}, {HOLD-REJECT}, {RETRIEVE}, {RETRIEVE-ACK}, {RETRIEVE-REJECT} in both directions.

ETS 300 764 describes the implementation of the Short Message service, point-to-point and cell broadcast.

The message handling and mapping for the support of the following GSM phase 2 services is included:

- SMS, point-to-point, Mobile Originated (SM MO)
- SMS, point-to-point, Mobile Terminated (SM MT)
- SMS Cell Broadcast (SMSCB)

ETS 300 756 covers the support of GSM bearer services. The specification describes the message handling and mapping of the following non-transparent GSM bearer services:

- Asynchronous (300, 1200, 1200/75, 2400, 4800, 9600 bit/s), GSM BS 21-26,
- Synchronous (1200, 2400, 4800, 9600 bit/s), GSM BS 31-34,
- PAD Access (300, 1200, 1200/75, 2400, 4800, 9600 bit/s), GSM BS 41-46.

Only GSM non-transparent (NT) bearer services are supported due to the usage of DECT air interface data profile service type C.2, which is based on a non-transparent protocol (LAPU).

The rates defined above, are the rates that are used by the application and the maximum rate is 9,6 kbit/s although the DECT data profile service type C.2 can provide in single slot operation up to 24 kbit/s rate on the air interface. The restrictions are due to the limitation of the GSM air-interface.

A further part of the GIP is **ETS 300 792** providing support of facsimile group 3. It contains the requirements and mappings necessary to ensure that the GSM facsimile group 3 service can be provided over the DECT air-interface. Transparent and non-transparent facsimile connections with the data rates of 2400, 4800 and 9600 bit/s are supported (GSM TS 62 automatic fax).

On the DECT air-interface the service is provided by the data profile F including the Multimedia Messaging Service Protocol, MMSP. The interworking shall take place on the ITU-T Recommendation T.30 protocol level. The DECT air interface U-Plane is based on the LAPU protocol error correction. The support of true transparent facsimile service can be supported by the data service profile D, class 2 and is for further study.

The maximum rate is 9,6 kbit/s due to the limitation of the GSM air-interface, even though the data service profile F can support a rate of 24 kbit/s in single slot operation.

2. DECT/GSM INTEGRATION

Why shall the two technologies DECT and GSM be connected to each other? Below we will outline some application scenarios and their feasibility, However, the profitable use or such a hybrid system very much depends on numerous circumstances like prices, market acceptance, terminal integration, availability, customer growth, the regulatory environment and competition.

2.1 Comparison DECT - GSM

To evaluate the potential benefits for a combination of GSM/DECT network elements it is helpful to compare main radio features. The major difference between the two standards is that GSM is a fully featured network specification, whereas DECT is an access standard which defines the interface between a mobile cordless terminal and a base station (only the lower three ISO/OSI layers are described). With the Dynamic Channel Selection / Allocation (DCS/DCA) of DECT no complex frequency planning is necessary. The handover procedures are controlled by the DECT terminal.

In the following some main criteria to distinguish GSM and DECT are shortly summarised:

	GSM	**DECT**
Frequency band	890 - 915 MHz 935 - 960 MHz	1880 - 1900 MHz
# Carriers	124	10
# Channels	992	120
Timeslots/carrier	8/16 (FR/HR)	12
Datarate/carrier	270.8 kbit/s	1 152 kbit/s
TDMA Frame duration	4.615 ms	10 ms
Power	8/5/2/0.8 W	0.25 W
Range	< 35 km	< 300 m and up to **3-5 km** with directive antennas ptmp.
Channel allocation	Network planning	dynamic, decentral
Interleaving	over 8 Frames	not foreseen
Handover-Control	central	decentral, Portable Part
System-capacity	200 Erlang/km²	10.000 Erlang/km²/floor
	10 Erlang/MHz/km²	500 Erlang/MHz/km²
Speech-quality	RPE-LPC-Fullrate (FR) Halfrate (HR) Enhanced Full Rate (EFR)	ADPCM
Datarate for speech	13 kbit/s (FR, EFR) 6,5 kbit/s (HR)	32 kbit/s

2.2 DECT in the cellular world

Where the future UMTS or IMT-2000 shall offer a global mobility throughout the world, GSM offers similar mobility but has some capacity constrains when addressing the residential market. Originally DECT was aiming at the cordless market. This target market was expanded by adding additional features to the DECT core standard, as described. With those features DECT is able to access the mass market for all residential customer and may be used in connection with wireless PBXs for the business customer.

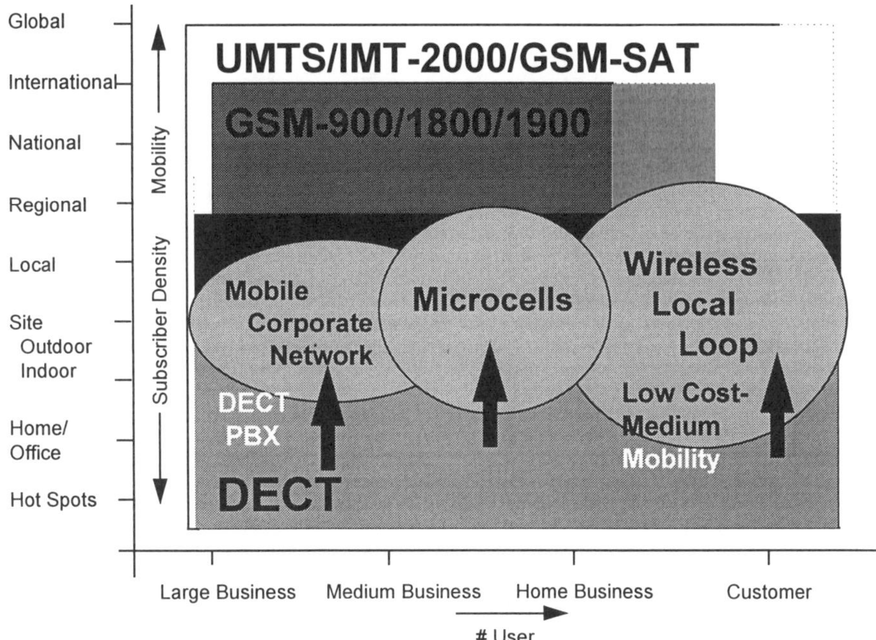

Figure 6: DECT in the cellular world

The combination of DECT and GSM is able to close the gap between cellular systems and the pure indoor systems in providing applications like mobile corporate networks (private networks), microcells, hot spots and low cost - medium mobility wireless local loop services.

2.3 Integration of DECT and GSM

The combination of DECT and GSM may let the operator use the best characteristics of each system and might be an invaluable alliance. The advantages of both systems potentially could be used in a DECT/GSM-hybrid solution to create a new dimension of mobile communication offering in-house coverage and mobility management and to help to be competitive with other systems and PCN-operators.

In high density areas, the radio infrastructure cost for DECT are supposed to be lower relative to GSM, that makes DECT an alternative to GSM in some environments.

The users may benefit from the mobility functions of GSM, offering a 'wide-area' mobility with DECT, or having access to GSM features with a high quality DECT radio link (i.e. speech or datarate). The operator, on the other hand, may reach new customers (DECT users) while using additional frequency spectrum.

2.4 DECT/GSM Scenarios

The following figure provides an overview of possible interworking scenarios and relevant applications.

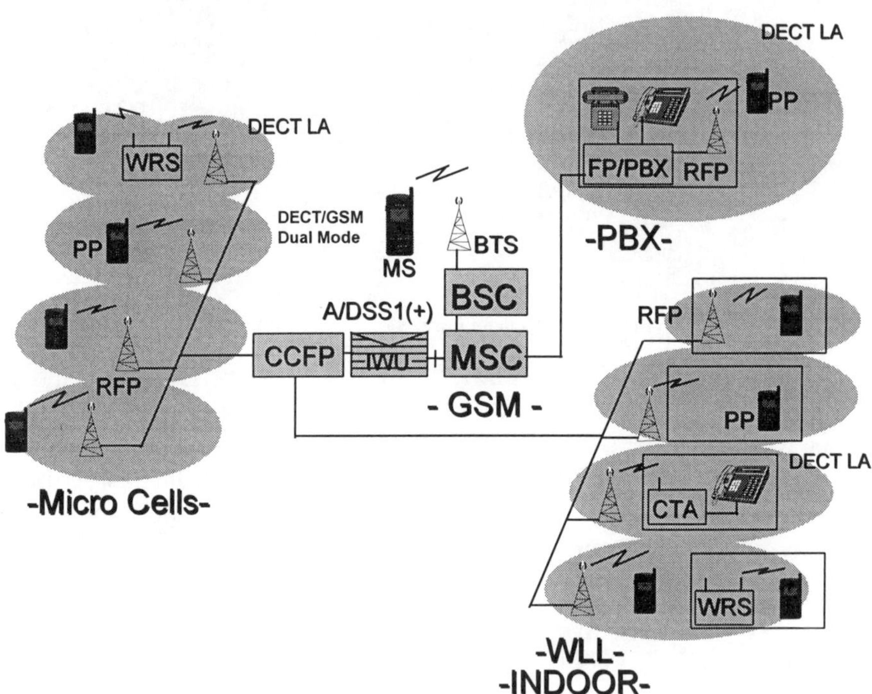

Figure 7: DECT scenarios

In the middle we see a part of the GSM network with one MSC, a BSC and a BTS. The MSC has a direct connection to a *PBX* or a *Private Integrated Services Network (PISN)*, using an ISDN based interface. The MSC A-interface connection with an InterWorking Unit (IWU) may be used for the connection to *micro cells*, providing DECT coverage in areas like cities or shopping centres. A centrex scenario is not described, as such services are not common in Europe, although possible as well. On the right side the connection of DECT to GSM offering *local loop services*, *indoor-* and *hot spots* coverage is shown.

2.5 DECT - GSM Connection

This chapter describes the interworking scenarios that are so far taken into account in several standardization groups.

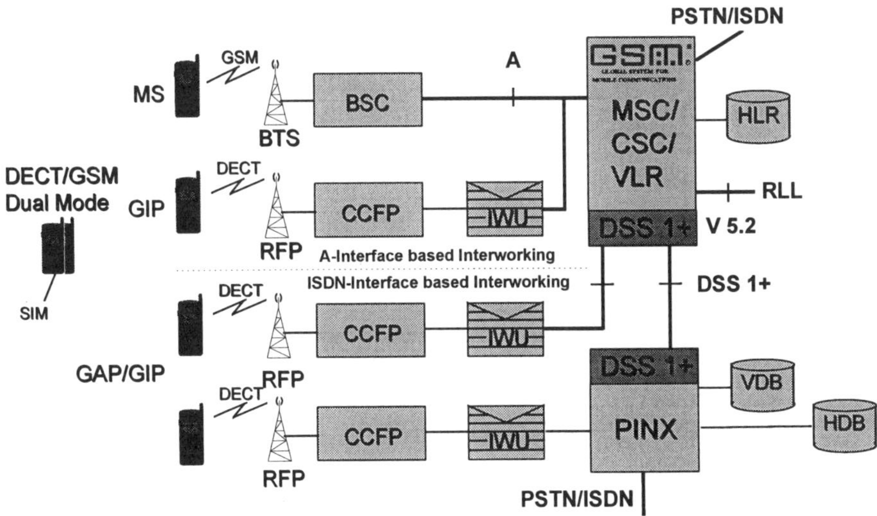

Figure 8: DECT network attachments

2.5.1 Access to GSM via the A-Interface

The A-interface connection of DECT to GSM, where the DECT part behaves like a 'BSS' for the GSM-MSC is the first scenario that was considered within ETSI. This scenario requires an interworking unit between the MSC and the DECT Fixed Part, that handles the protocol mapping and conversion of the A-Interface messages to DECT protocol messages and vice versa. One major advantage of this solution is that there are no changes necessary on the GSM side. In the years 1994-1997 ETSI prepared a set of standards that are now summarised as the DECT **G**SM **I**nterworking **P**rofile GIP. Detailed explanations of the different standards can be found in chapter 1.2.4.

2.5.2 Access to GSM via the DSS1+ -Interface

This scenario is pretty much the same as the previous one, but the DSS1+-protocol is assumed to be the interface to GSM.

GSM core services are provided to the users of cordless DECT terminals with GSM subscription, via co-operating GSM and DECT network elements which are interconnected by means of an ISDN interface. The DECT network elements may or may not be under the authority of a GSM network provider. ETS 300-787 and -788 'DECT access to GSM via ISDN' stage 1 and stage 2 describe the interworking.

2.5.3 DECT/GSM Network-to-Network Interworking (DSS1+)

In this scenario the direct connection of a DECT-PISN to the GSM MSC may offer an attractive opportunity. The DECT network elements consist of PINX equipment, that is able to control DECT specific mobility and DECT intra-cell and intra-cluster handover.

Providing local mobility (using DECT) and global mobility (using GSM) with one number and one common set of services is the major advantage for the user. With the possibility to support roaming between DECT islands, supported by the exchange of mobility information between the DECT-PINX and the GSM network (HLR/VLR), the support of a private numbering plan and the usage of DECT/GSM dual-mode terminals, an integrated and highly sophisticated seamless service could be offered.

Only the A-interface and the DSS1+, that is actually under elaboration, offer the possibility to exchange the relevant mobility management information. Common PBXs don't support the complex SS#7 protocol but the ISDN interfaces, therefore the DSS1+ protocol is getting more and more important.

The PISN is to be regarded on a *peer-to-peer level* as a different network and as a prerequisite commercial agreements between the GSM network and the private DECT network must be established. However, contrary to GSM, the PISN is seen as a single visited area. The support of the relevant security mechanisms and the appropriate mapping has to be ensured, e.g. mapping of IMEI - IPEI, IMSI - IPUI-R, TMSI - TPUI,

Roaming of users of cordless DECT terminals between private DECT networks and public GSM networks cover two basic scenarios:

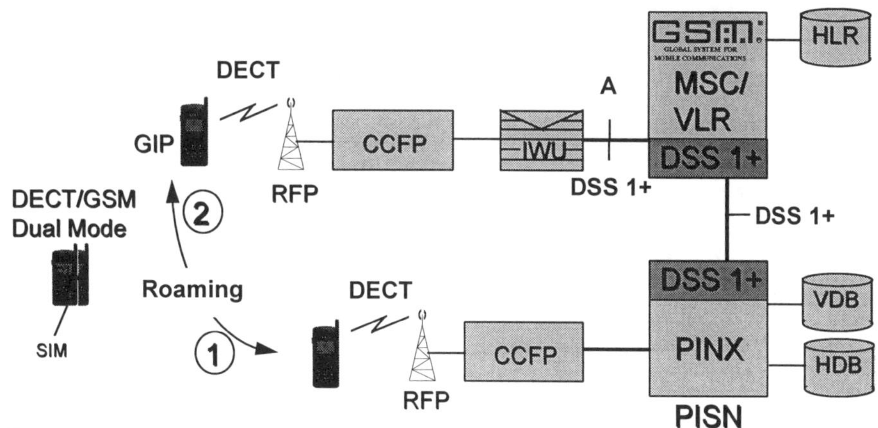

Figure 9: DECT - GSM roaming

1. The cordless DECT terminal user is a GSM subscriber, roaming to a private DECT network

2.The cordless DECT terminal user is a user of a private DECT network roaming to a GSM network.

2.5.3.1 Relation to Cordless Terminal Mobility (CTM)

Cordless Terminal Mobility, CTM, include network based mobility functions based on ISDN and intelligent network (IN) facilities offering roaming between CTM networks.

In contrast to the CTM approach the DECT/GSM interworking concept would use GSM inherent mobility functionality while using the IN platform for service differentiation.

CTM does not intend to provide the mobility that is provided by GSM, and might be seen as complementary to GSM.

2.5.4 DECT/GSM Dual-mode Terminals

DECT/GSM dual mode terminals combine the cordless and cellular technology in one casing, using the same MMI, speaker, microphone, display and so on.

With some intelligence in the network, a DECT/GSM interworking, purely based on a DECT/GSM dual mode terminal, is possible.

First prototypes were made available already in 1995, the availability of commercial products depends on demand.

Meanwhile an ETSI Technical Report describe 5 different types of DECT/GSM dual mode terminals:

Table 2. DECT/GSM dual mode terminal types

Type of Terminal	Description
1	Manual switching, active in one mode at a time
2	Automatic switching, active in one mode at a time
3	Automatic switching, listen in both modes
4	Automatic switching, listen in one mode, active in the other mode
5	Automatic switching, active in both modes possible

A type 2 terminal seems to be realistic, could be made available at reasonable costs today and works with existing network infrastructures. During switching from one mode to the other, the terminal is unable to make or receive calls for several seconds.

2.6 Further Developments

2.6.1 Enhanced Bearer Services

The used protocols and rate adaptations for data services in GSM are strongly correlated to the fact that there is the bottleneck of the GSM air-interface with its 9.6/14,4 kbit/s user-bandwidth and the necessity to provide a secure link even in a difficult radio environment. The DECT air-interface allows 32 kbit/s per channel or even up to 552 kbit/s using the DECT multi bearer service. Utilisation of the high capacity DECT air-interface in connection with GSM would lead to a much better service performance. For the support of DECT data services with a bandwidth up to 64 kbit/s nomal A-interface PCM link switching is applicable and data can be passed transparently through the MSC without using interworking functions and modems.

The work on this item was started during the year 1996. It is closely related to future evolutions in GSM like High Speed Circuit Switched Data (HSCSD) and the General Radio Packet Service (GPRS).

Two new work items for the support of DECT/GSM interworking to HSCSD and GPRS were initiated by ETSI's DECT Project.

GPRS as a part of the GSM phase 2+, offers packet oriented bearer services, point-to-point and point-to-multipoint. GPRS data rates over the GSM air-interface will vary from about 9,05 kbit/s up to 21,4 kbit/s x 8 = 171,2 kbit/s.

For the realisation of DECT/GPRS interworking, the DECT system will be attached to GPRS via the Gb interface, this is the interface between the Serving GPRS Support Node (SGSN) and the Base Station Subsystem (BSS). From the SGSN point of view the DECT interworking unit (IWU) looks like a ordinary BSS.

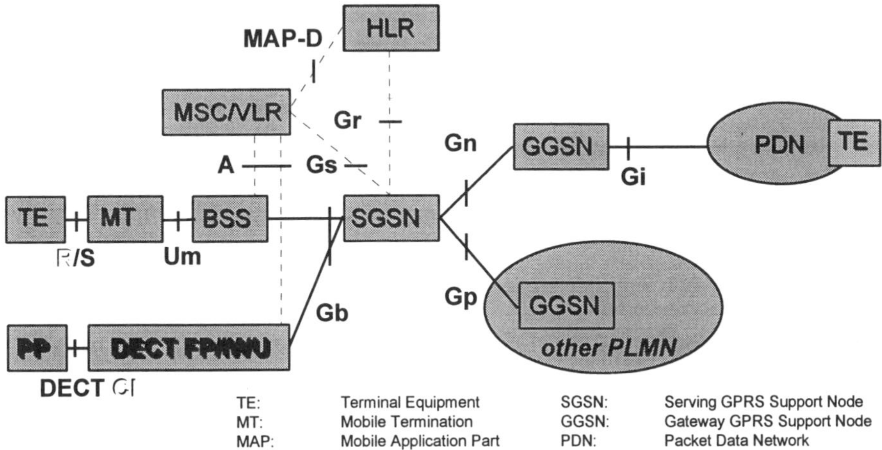

Figure 10: Possible DECT-GPRS Architecture

The DECT interworking unit has to be equipped with a Frame Relay interface that is also used on the Gb interface.

The DECT air-interface data profiles B.2 and C.2 are optimised for packet data, offer data rates up to 552 kbit/s and are suitable for the support of GPRS.

On the terminal side, the DECT portable part will be equipped with interworking functions that translate the DECT specific protocols to GPRS protocols. It has to be investigated, how much of the existing DECT/GSM Interworking specification GIP may be reused in DECT/GPRS interworking.

3.. DECT/GSM APPLICATIONS - EXPERIENCES

3.1 Residential

In the residential market DECT has gained an enormous success by reaching already a share of about 15% in the year 1996.

Convergence of fixed and mobile applications is one of the key requirements for 3rd Generation systems. The described DECT/GSM interworking capabilities may contribute to this goal even for 2nd generation cellular systems.

Today, the DECT systems are usually attached to the PSTN/ISDN. In the future GSM might be used as the intelligent backbone providing all features necessary to compete in the market. The add-on will be the enhanced mobility for the customers, realised e.g. by reusing already existing copper-cables of a fixed operator and providing mobility management by using the DSS1+ interface as described in chapters 2.5.2 and 2.5.3.

If the fixed access to the customer isn't feasible, the DECT wireless local loop application described in chapter 3.1.1 might be a solution.

3.1.1 Radio in the Local Loop

The average level of telephone penetration outside of Western Europe and North America continues to be less than 5 per cent. The main obstacle lies in installing the massive number of lines required in the local loop. Traditional methods of laying copper cable pairs to each subscriber are slow and costly and operators are increasingly turning to radio as a substitute for copper in the local loop. DECT has a number of advantages when deployed instead of copper cable in the local loop.

For some countries offering just pure telephony (POTS) will not satisfy customers' demand. Facsimile, fast modem compatibility and 64 kbit/s ISDN services need to be offered as well.

For that reason ETSI has developed the Radio in the local loop Access Profile (RAP) and the ISDN interworking profiles to cover this kind of application with DECT fixed radio access systems.

Such Radio local loop systems based on DECT have already been installed in several countries, but are not connected to the GSM network so far, although possible.

Trials or commercially running DECT RLL systems can be found in Bahamas, Bahrain, Bolivia, Brazil, Cambodia, Chile, China, Colombia, Czech Republic, Denmark, Finland, France, Germany, Greece, Hungary, India, Indonesia, Mexico, Myanmar, Namibia, Norway, Philippines, Poland, Romania, Slovak Republic, Singapore, South Africa, Spain, Sri Lanka, Sweden, Switzerland, Uruguay and Venezuela.

3.2 Business

Major demand for cordless systems will be driven by the office market - a European Commission estimate suggests that as many as 30 per cent of all office telephones behind a PBX in Europe will be cordless by the year 2000.

For the small, medium and large business sectors DECT is now established as the favourite cordless technology.

DECT systems attached to PBXs can easily be connected to a GSM network using an ISDN based interface that is common in all PBXs.

When offering roaming between GSM and DECT-PBXs or private networks and when enabling GSM subscribers to access their GSM services using the DECT system, the upgrade to the extended ISDN interface, DSS1+, is necessary. The technical solution is described in chapter 2.5.3. Also DECT/GSM dual mode terminals enable the interworking between the business and the cellular environment and represent an important element for the integration.

Although standards for connecting GSM to private networks are currently in preparation, still various questions need to be addressed:

Do we need thousands of roaming- or commercial-agreements with the PBXs or private networks, how can we support home and visited services when roaming into the visited environment, what about charging aspects and the usage of the home numbering plan...

3.3 Public

3.3.1 Public Pedestrian and Hot Spots

The open character of DECT standards enables DECT to be expanded to public services in areas such as city centres, campuses and airports, covering public pedestrian and hot spots applications.

When having GSM as the network backbone, the DECT access network supports the mobility within one service zone while the GSM mobility management supports the wide area mobility between different zones. The driving force is usually a meaningful capacity enhancement of the existing network by introducing the alternative air-interface DECT.

Technically feasible solutions for the interworking are both the A-Interface and the ISDN-interface based interworking described in chapter 2.5.

Obviously there seems to be a reluctance in the market to introduce public DECT/GSM interworking services, either for enhancing the capacity of GSM in certain areas, e.g. Hot Spots, or offering an integrated wireless local loop service including public pedestrian mobility. What are the reasons?

- Offering a public pedestrian DECT service (CTM) with a city-wide coverage is very expensive due to intensive planning of antenna-configurations, the high amount of basestations and the enormous cabling costs. In summary, the smaller the radio cells the higher the investment in the fixed network. Deploying a full coverage, high capacity micro cellular network using DECT might be as expensive as a similar fixed network.

In Italy a CTM service based on the fixed network and IN, using more than 100.000 basestations, will offer urban wide mobility including roaming between cities. Such a system might be economically useful, if the usage of existing cables is possible.

- With multiple DECT operators offering their services in the same geographical area in the unlicensed band of 1880-1900 MHz, potential DECT/GSM operators are not able to ensure a guaranteed level of Quality of Service to the customer due to possible influences from other DECT systems.

- The uncertainty of the National regulatory environment might be a hurdle for potential operators.

It seems doubtful whether the introduction of a DECT/GSM public pedestrian service really reaches a new market segment, or whether also GSM can provide cost effective and sufficient access for this environment.

GSM operator still have the possibility to use some of the specified GSM features to enhance the GSM capacity and quality of the network in a cost efficient way, e.g.:

- Frequency hopping, Voice Activity Detection (VAD), Discontinuous Transmission (DTX), Power Control
- Micro BTS, Pico Cells

- Smart antennas
- GSM cordless telephony system (home basestation)
- Half Rate Codec, Adaptive Multi Rate Codec (AMR)
- GSM 900 and 1800 MHz frequencies

4. TIMESCALE OF STANDARDS

4.1.1 DECT basic standard/documents

Table 3. DECT basic documents

Reference	Name/Description	Status/Time Schedule
ETS 300 175-1-9	DECT standard Edition 2/3	published 09/96
ETS 300 825	3 Volt DECT Authentication Module (DAM)	published 09/97
ETS 300 329	EMC standard for DECT equipment	published 09/94
EN 300 444	DECT Generic Access Profile (GAP)	published 12/95
EN 300 824	CTM Access Profile (CAP)	published 10/97
ETS 300 700	DECT Wireless Relay Station (WRS)	published 03/97
DEC/DECT-050116	Optional modulation schemes for services up to 2 Mbit/s	1999*

4.1.2 DECT DATA standards

Table 4. DECT DATA standards (Class 2)

Reference	Name/Description	Status/Time Schedule
ETS 300 701	Service Types A and B, Class 2, Data Services Profile, Generic frame relay service with mobility	published 10/96
EN 300 651	Service Type C, Class 2, Data Services Profile, Generic data link service	published 09/96
EN 301 238	Data Services Profile (DSP); Isochronous data bearer services with roaming mobility (service type D, mobility class 2)	1998*
ETS 300 757	Data Services Profile (DSP); Low rate messaging service; (Service type E, class 2)	published 04/97
ETS 300 755	Data Services Profile (DSP); Multimedia Messaging Service (MMS) with specific provision for facsimile services; (Service type F, class 2)	published 05/97
DE/DECT-020087	DECT Data Service Change (data multimedia)	started

EN 301 240	Data Services Profile (DSP); Point-to-Point Protocol (PPP) interworking for internet access and general multi-protocol datagram transport	1998*

4.1.3 GIP - DECT/GSM Interworking documents

Table 5. GIP - DECT/GSM Interworking documents

Reference	Name/Description	Status/Time Schedule
ETS 300 370	DECT/GSM Interworking Profile (IWP); Access and mapping (protocol/procedure description for 3,1 kHz speech service), Edition 2	1998*
ETS 300 466	General description of service requirements, functional capabilities and information flows	published 07/96
ETS 300 499	GSM-MSC—DECT-FP Fixed interconnection	published 09/96
EN 300 703	DECT/GSM Interworking Profile (IWP); GSM phase 2 supplementary services implementation	1998*
ETS 300 764	DECT/GSM Interworking Profile (IWP); Implementation of short message service, point-to-point and cell broadcast	published 05/97
ETS 300 756	DECT/GSM Interworking Profile (IWP); Implementation of bearer services	published 03/97
ETS 300 792	DECT/GSM Interworking Profile (IWP); Implementation of facsimile group 3	published 06/97
EN 301 242	Digital Enhanced Cordless Telecommunications (DECT); Global System for Mobile communications (GSM); DECT/GSM integration based on dual-mode terminals	1998*
DEN/DECT-010079	DECT/GSM Interworking Profile (IWP); Enhanced bearer services	1998*
DEN/DECT-010097	DECT/GSM Interworking Profile (IWP); Implementation of General Packet Radio Service (GPRS)	1999*
DEN/DECT-010098	DECT/GSM Interworking Profile (IWP); Implementation of High Speed Circuit Switched Data (HSD)	1999*

4.1.4 DECT access to GSM via DSS 1 +

Table 6. DECT access to GSM via DSS 1+

Reference	Name/Description	Status/Time Schedule

ETS 300 787	DECT access to GSM via ISDN; General description of service requirements	published 07/97
ETS 300 788	DECT access to GSM via ISDN; Functional capabilities and information flows	published 07/97
DEN/DECT-010065	DECT/GSM Interworking Profile (IWP); DECT access to GSM Public Land Mobile Network (PLMN) via ISDN+ interface; Bearer services and Short Message Service (SMS)	1999*
DEN/DECT-010066	DECT/GSM Interworking Profile (IWP); DECT access to GSM Public Land Mobile Network (PLMN) via ISDN+ interface	1999*
DEN/DECT-010067	DECT/GSM Interworking Profile (IWP); DECT access to GSM Public Land Mobile Network (PLMN) via ISDN+ interface; GSM supplementary services	1999*

4.1.5 DECT ETSI Technical Reports

Table 7. DECT ETSI Technical Reports

ETR 015	DECT reference document	published 03/91
ETR 041	DECT, transmission aspects, 3.1 kHz telephony	published 07/92
ETR 042	Guide to features that influence traffic capacity	published 07/92
ETR 043	DECT services and facilities	published 07/92
ETR 056	DECT system description document	published 07/93
ETR 139	Radio in the Local Loop (RLL)	published 11/94
ETR 183	Conformance test specification for DECT	published 11/95
ETR 159	DECT wide area mobility services using GSM	published 07/95
ETR 178	A high level guide to the DECT standardization, Edition 2	published 01/97
ETR 185	Data services profile, overview	published 12/95
ETR 246	Wireless Relay Stations (WRS)	published 11/95
ETR 341	DECT/GSM IWP Profile Overview	published 01/97
ETR 308	Radio local loop Access Profile (RAP)	published 08/96
ETR 310	Traffic capacity and spectrum requirements for multi-system & multi-service applications co-existing in a common frequency band	published 08/96
TR 101 072	Integration based on dual-mode terminals	published 06/97
DTR/DECT-010096	DECT/GSM advanced integration of dual-mode terminals	1998*

4.1.6 DECT ETSI Technical Basis for Regulation

Table 8. DECT ETSI Technical Basis for Regulation

TBR 006 Ed. 2	General terminal attachment req. (second edition)	adopted
TBR 010 Ed. 2	General terminal attachment req., telephony applications (second edition)	adopted
TBR 011	PAP attachment for terminal equipment	published 09/94
TBR 022	GAP attachment for terminal equipment	adopted
TBR 036	DECT access to GSM Public Land Mobile Networks (PLMNs) for 3,1 kHz speech applications	1998*
TBR 039	DECT/GSM dual-mode terminals	1998/99*

*expected publishing date

REFERENCES

- all above mentioned standards and draft Recommendations
- DECT Forum: www.dect.ch
- ETSI: www.etsi.fr
- Cordless Telecommunications Worldwide - Walter H.W. Tuttlebee (Ed.), 1997
- Presentation held at EUROFORUM DECT/GSM Conference in London, 1997
- Presentation held at IIR DECT/GSM Conference in London, 1997

ABBREVIATIONS

ADPCM	Adaptive Differential Pulse Code Modulation
ARI	Access Rights Identity
ARQ	Automatic Repeat reQuest
BSC	Base Station Controller
BSS	Base Station System
BTC	Business TeleCommunications
BTS	Base Transceiver Station
CAP	CTM Access Profile
CC	Call Control. A NWK layer functional grouping
CCFP	Central Control Fixed Part
CCITT	International Telegraph and Telephone Consultative Committee

CDMA	Code Division Multiple Access
CEPT	Conference of European Posts and Telecommunications
CI	Common Interface (standard)
CK	Cipher Key
CODEC	COder-DECoder
CRC	Cyclic Redundancy Check. A cyclically generated field of parity bits
CRFP	Cordless Radio Fixed Part
CSF	Cell Site Function. A MAC layer functional grouping
CSPDN	Circuit Switched Public Data Network
CTM	Cordless Terminal Mobility
CTA	Cordless Terminal Adapter
C-PLANE	Control PLANE
DAM	DECT Authentication Module
DCK	Derived Cipher Key
DECT	Digital European Cordless Telecommunications
ES	End System
ETSI	European Telecommunications Standards Institute
FP	Fixed Part
FT	Fixed radio Termination
GAP	Generic Access Profile
GFSK	Gaussian Frequency Shift Keying
GGSN	Gateway GPRS Support Node
GIP	DECT GSM Interworking Profile
GNW	Global NetWork
GPRS	General Packet Radio Service
HDB	Home Data Base
HLR	Home Location Register
HSCSD	High Speed Circuit Switched Data
IA5	International Alphabet No.5 (defined by CCITT)
IFEI	International Fixed Equipment Identity
IN	Intelligent Network
INAP	IN Application Part
IP	Internet Protocol
IPEI	International Portable Equipment Identity
IPUI	International Portable User Identity.
IWU	InterWorking Unit
ISDN	Integrated Services Digital Network
IWF	InterWorking Functions
IWU	InterWorking Unit

K	authentication Key
LAPC	DLC layer C-plane protocol entity
LAPU	DLC layer U-plane protocol entity
LAN	Local Area Network
LCE	Link Control Entity
MAC	Medium Access Control
MBC	Multiple Bearer Control
MCEI	MAC Connection Endpoint Identification
MM	Mobility Management
MMSP	Multi Media Messaging Service Profile
MSC	Mobile Switching Centre
MUX	time MUltipleXor
NWK	NetWorK. Layer 3 of the DECT protocol stack
O&M	Operations & Maintanance
OSI	Open Systems Interconnection
PA	Portable Application
PARI	Primary Access Rights Identity
PARK	Portable Access Rights Key, states the access rights for a PP
PBX(PABX)	Private Automatic Branch eXchange
PDU	Protocol Data Unit
PHL	PHysical Layer 1 of the DECT protocol stack
PINX	Private Integrated services Network eXchange
PISN	Private Integrated Services Networks
PLMN	Public Land Mobile Network
POT(S)	Plain Old Telephone (Service)
PP	Portable Part
PSPDN	Packet Switched Public Data Network
PSTN	Public Switched Telephone Network
PT	Portable radio Termination
PTN	Private Telecommunication Network
PUN	Portable User Number
PUT	Portable User Type
PWT	Personal Wireless Telecommunications
REP	REpeater Part
RF	Radio Frequency
RFP	Radio Fixed Part
RFPI	Radio Fixed Part Identity
RPN	Radio fixed Part Number
RSSI	Radio Signal Strength Indicator

SARI	Secondary Access Rights Identity
SCP	Service Control Point
SDP	Service Data Point
SDU	Service Data Unit
SGSN	Serving GPRS Support Node
SMG	Special Mobile Group
SM MO	Short Message Mobile Originated
SM MT	Short Message Mobile Terminated
SMS	Short Message Service
SMSCB	SMS Cell Broadcast
SS	Supplementary Services
TARI	Tertiary Access Rights Identity
TBR	Technical Basis for Regulation
TDD	Time Division Duplex
TDMA	Time Division Multiple Access
TI	Transaction Identifier
TPUI	Temporary Portable User Identity
UAK	User Authentication Key
UPI	User Personal Identification
U-PLANE	User PLANE
VDB	Visitors Data Base
VLR	Visitor Location Register
WRS	Wireless relay station

PART 4

PROGRESS TOWARDS UMTS

Chapter 1

ATDMA

Efficient Access for Tomorrow's Mobile Communication

N. Metzner
Siemens AG ÖN MN

Key words: UMTS, third generation wireless systems, RACE, TDMA.

Abstract: The plan for the future of mobile communications is to merge previously separate systems for voice and data transmission (cellular mobile radio, cordless telephones, and paging systems) into a universal mobile communication system (UMTS) with an extended range of services for voice and data, variable bit rates, and compact mobile terminals. The development of this third generation mobile radio system focuses on spectrum-efficient radio access technology. In Europe investigation and development in this field is supported by the Commission of the European Union, currently within the ACTS program, previously within the RACE program. Thus the RACE project ATDMA investigated the potential of advanced TDMA systems with respect to the goals of UMTS. Propagation measurements, channel modelling and techniques studies were performed, a real-time and simulated testbed were built to prove feasibility and efficiency of this approach.

This contribution gives an overview about the project's content and its main results.

1. INTRODUCTION

In the third generation of mobile radio systems (e.g. UMTS and FPLMTS), the numerous voice and non-voice services with variable bit rates are to be combined in one common system, offering users better service, more convenience, and greater mobility. Within the RACE program, the two projects ATDMA and CODIT were concerned with

TDMA or CDMA access methods, examining them from the following viewpoints: support of different bit rates, increase of user capacity for a specific traffic model, and analysis of the potential of different access technologies and how they might be improved.

To achieve the UMTS objectives, an adaptive air interface with the properties mentioned above is to be employed. Future mobile radio systems will operate with different types of radio cells; three are envisaged for UMTS, designated pico, micro, and macro. The wide range of conditions in these cells and the transition permitted between these transmission scenarios should be taken into account with this air interface.

2. ATDMA'S AREAS OF INTEREST

Within ATDMA the work was progressed simultaneously on four technical areas [1]:

2.1 Channel Characterisation

The mobile radio channel is characterised by multipath propagation, shadowing, and the Doppler effect caused by moving transmitters and receivers. This results in a time-variant channel impulse response or a transfer function with frequency selective fading. To guarantee the availability, reliability, and required quality of a mobile radio system, while also improving the spectrum efficiency, the channel characteristics corresponding to the access method must be taken into account. So far propagation data contained in the technical literature generally only applied for the 900 MHz frequency band. For this reason, the ATDMA project examined very extensively the radio channel in the UMTS frequency band at around 2 GHz with the aid of channel sounders based on correlation techniques and optimum estimation.

Broadband propagation was measured in all three radio cell types - pico (radius up to 100m), micro (100m to 1km), and macro (1km to 20km). In addition, heterogeneous scenarios (transitions from one cell type to the other) were examined.

From the propagation measurements the statistics of channel parameters (e.g. delay spread, delay window, interference power ratio, coherence bandwidth, correlation time, and path loss) were determined, which delivered the basic figures for system development. These included equaliser length, computing power for the signal processing algorithms, interleaving depth, and coding schemes.

2.2 Techniques Studies

In ATDMA, a distinction was made between the *transport* and *control* functions of a UMTS access system. If the adaptive access method is in a stable condition, then the control functions monitor the execution of the transport functions. As soon as disturbances affect this condition, adaptation functions are activated to once again shift the radio connection into a stable condition with sufficient quality.

This approach had been described as a supplement to the conventional OSI reference model in a *functional radio access model* (Fig. 1) [2]. It handles the details hierarchically, each level representing a different process of the system. Most of these function groups consist of several function units and are distributed among the various physical units. The air interface that should be adaptive for users, operators, cell type, service traffic load, and connection was developed according to this model. The focus was on the transmission plane and the lower layers of the control plane of the functional model. Some areas (e.g. the voice encoding algorithm) were not subject of that project.

Transport: The individual elements of the transmit and receive chain were examined for user and control data. The aim was to develop a suitable combination of modulation, coding, and interleaving techniques as well as block lenghts. In addition, those studies took into account control mechanisms (e.g. ARQ procedures) in the transport chain as a basis for the adaptation of the ATDMA air interface.

Control encompassed mechanisms to monitor and adapt the performance of the transport functions and to maintain link quality. Algorithms - from the adaptive control of the air interface via the handover to the PRMA, and from the dynamic channel assignment (DCA) to the dynamic resource allocation (DRA) - were developed.

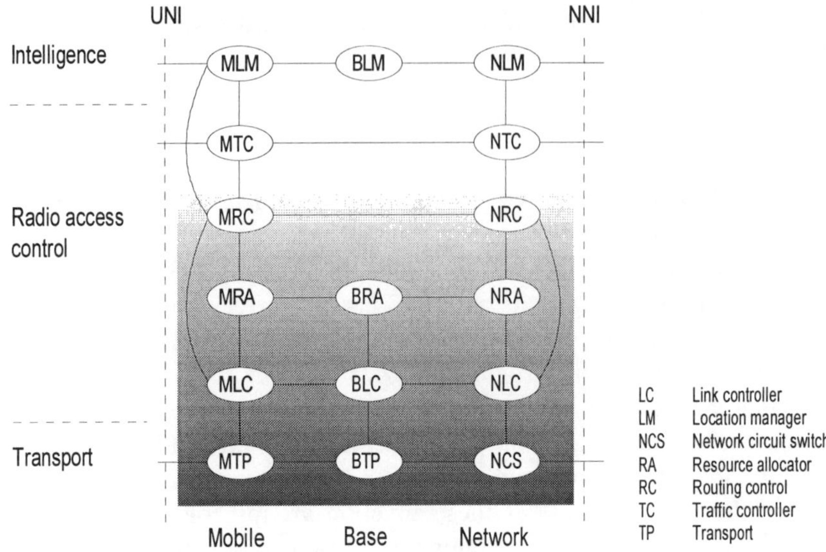

Fig. 1 Functional Radio Access Model developed within RACE

Testbed Implementation

To investigate system performance capability, the ATDMA project developed two test systems [3]: while a *real-time testbed (RTTB)* mainly permitted the checking of transport or of the lower control technologies under realistic conditions, a *simulation testbed (STB)* with software tools was used to examine more complex technologies.

In this way, the two testbeds complemented each other with regard to the implementation of algorithms. They differed in the degree of resolution and the time dimensions considered.

The RTTB (Fig. 2) consists of a powerful hardware platform for baseband processing, corresponding RF hardware, and a special equipment for functions specific to the interface and test system. The system configuration comprises five units (two base stations, two mobile stations, and a network emulator) and is implemented with a number of multiple-link digital signal processors (DSPs) to insure adequate processing power while retaining high flexibility. The purpose of the simulation testbed was to supply detailed information about the increase in system capacity brought about by the new ATDMA technologies. Of particular interest was the determination of the signalling load caused by these procedures. Due to the long time spans of control algorithms and the resulting computation load, a distinction was made between two types of simulation testbed:

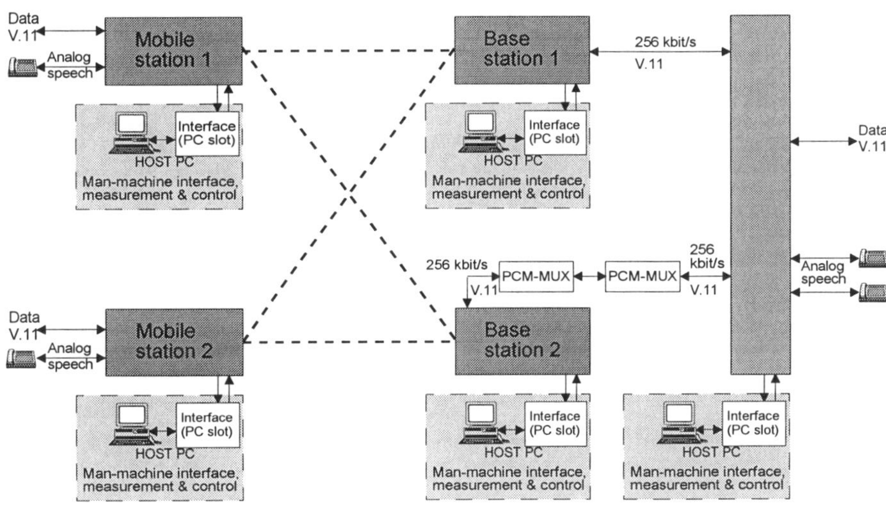

Fig. 2 Real-time Testbed developed within ATDMA

- The *detailed simulation testbed (D-STB)* describes processes from the burst level to the handover cycle in a limited geographical area and one uniform cell type, allowing detailed simulations at burst level to study the fine behaviour of control functions and their interactions.
- The *global simulation testbed (G-STB)* models processes in several cell types and environments in a larger geographical area with a lower time resolution to study cellular aspects.

2.3 System Performance Evaluation

Common criteria were devised for the evaluation and performance assessment of the solutions developed in the ADTMA and CODIT projects, and for the comparison of the two technologies with regard to system capacity, minimum required signal/noise ratios and signal/interference-ratios, network complexity, infrastructure requirements, and signalling load. Main objective of testing within ATDMA was to verify the project's system concept against the ATDMA project goals and high level objectives. These high level objectives were related to *system* - system capacity and complexity - and *technique evaluation* - individual technique performance, interaction of techniques - and were valid for all test phases. The tests consisted of validation tests with the STBs, real-time tests under reproducible conditions, and real-time field tests in the pico, micro, and macrocells with the RTTB.

3. THE PROJECT'S AIR INTERFACE PROPOSAL

Very low BER requirements for data services despite continuously varying propagation and interference conditions - it may not be efficient to design a fixed air interface but rather an adaptable air interface able to cope with these conditions. Following the project's philosophy techniques for *transport* and *control* were developed, supported by a control channel structure.

3.1 Transport Techniques

Development started with the design of the Burst, Frame and Multiframe Structure and resulted in a common frame length of 5 ms with all carrier bit rates derived from one single clock (Table 2). One of the selection cri-teria for the Modulation schemes in ATDMA was the robustness to power amplifier non-linearity. Binary Offset-QAM, which is nearly constant envelope modulation concept, was selected as the common basic modulation scheme for most cell types. For large macrocells GMSK, a constant envelope scheme will be used. Thus both schemes can be detected by the same receiver, i.e. receiver complexity will not be increased. In small cells where the transmission powers are rather low and non-linearity of the amplifier can be easier handled higher level quarternary QAM can be used to increase bit rate.

This was one feature on the way to the design of *Transport Modes*, a „toolbox“ from which the actual mode of each element can be selected according the propagation environment and the service requirements. The resulting different operating modes for the air-interface are a set of {modulation alphabet, channel coding, interleaving depth, radio resources}, able to guarantee a given quality at a given signal-to-noise plus interference ratio [2]. The commonalties between the different transport modes, simplifying the structure of the transceiver, are

- the same baseband receiver can be used basically in all operation modes,
- one common clock for all frequencies required,
- only two different IF band

In summary only the physical layer is operation mode dependent.

Conventional error protection meets the requirements for the non-corruption of data in almost every situation but requires extensive resources. An adaptive air interface meeting the same protection requirements provides only those resources that are absolutely necessary, thereby achieving a significant improvement in spectrum efficiency. Apart from the variable error protection, this results in advantages for other radio resources such as transmit power, mo-dulation stage, and interleaving. Thus Coding played a key role within the transport modes. ATDMA designed error protections mechanisms for speech services, low delay data services, long delay data services, and unconstrained delay data services. For example, for low and long constrained data a concatenated coding scheme was specified, with a fixed inner convolutional coder and a variable (through code shortening) outer Reed Solomon code. Table 1 lists the modes for the long constrained delay data service at 64 kbit/s with a delay constraint of 300ms, interleaved on 36 frames.

mode	inner code rate	outer code rate	global rate	coded bits per block	slots / frame
1	1 / 2	1	1 / 2	3960	10
2	1 / 2	3 / 4	3 / 8	5544	14
3	1 / 2	1 / 2	1 / 4	7920	20

Table 1 Transport mode of the long constrained delay data service at 64 kbit/s. Please note that mode 3 cannot be supported on a single carrier in macrocells.

The Equalisation was not part of the ATDMA research. However ATDMA showed that feasible algorithms exist with acceptable complexity, e.g. DFE and DFSE. The complexity of the 16 stage DFSE is about twice the complexity of a GSM equaliser.

Measurement Methods were defined to enable proper adaptation of the transport chain by the control techniques.

3.2 Control Techniques

Adaptive Power Control controls the transmit power on both up- and downlink. The control algorithm uses link quality measurements on the link being controlled for longer term control, and measurements on the reverse link for short-term control. The range of short-term control is controlled by the BS and adjusted according the pathloss difference observed between up- and downlink. By keeping the transmit power as low as possible power control reduces co-channel interference and thus reduces the frequency re-use factor. This is especially important regarding the high target Quality of Service (QoS) and coverage probabilities envisaged for UMTS.

The Handover process - a mobile station based, self-adapting algorithm - has three main phases: measurements, decision and execution with the measurement and the decision phases actually being continuous processes operating in parallel. Since propagation conditions are likely to be changing too rapidly it is assumed that all normal handover trigger decisions are taken by the mobile. When designing this procedure a trade-off between minimising the number of handover performed, for signalling purposes, and keeping an acceptable QoS for all calls had to be found. This algorithm will enable fast, seamless and reliable handover. Furthermore, a macrodiversity scheme was considered in ATDMA for micro cellular environment. It is not absolutely required but may improve the network efficiency at the expense of complexity.

Packet Access allocates capacity on de-mand using a technique called PRMA++ [4], a protocol which avoids wasting capacity during breaks in traffic source activity, or the need to permanently allocate capacity for the maximum bit rate a call may need. PRMA++ is centrally controlled and is based on separate access control and traffic channel resource allocations.

Based on the need to ensure that service quality is being maintained with a minimum amount of radio resources Link Adaptation - a radio access control technique - selects the operating transport mode. Transport mode (set of coding scheme, modulation, interleaving, etc.) changes are performed whenever average channel quality or transmit power are observed to be outside their permitted operation ranges.

Dynamic Channel Allocation seems to pro-mise an efficient handling of time and location variant traffic load in mobile communication systems. Furthermore, DCA will support an easy way of introducing new base stations into a network when calibrated carefully. The studied scheme allows a BS to use all frequencies achieving channel segregation by monitoring time-slot quality and assigning them to TCH's according a preference list (based on interference observed) in the base station.

3.3 Control Channel Structure

The ATDMA concept foresees to reserve a given number of slots per TDMA frame for signalling purposes. However the signalling channels requirements depend on the frame structure and the traffic load. The signalling channels are transmitted on a multi-frame basis, which consists of 128 TDMA frames (640ms). The channels designed were:

LCCH: a novelty, developed by ATDMA; one slot per multiframe for each call carried in the cell for mobility management.

ACCH: uni-directional: forward (direction is related to the TCH): burst internal signalling, return: 8 slots per multiframe for each traffic slot assigned to a link.

CCCH: The requirements mainly depend on PRMA++ optimisation; 2 Reservation-slots (R-slots) are required for macrocell and 4 R-slots are required for micro and picocell (single carrier per cell configuration). Trunking efficiency will increase and thus signalling requirements will be relaxed for the multi-carriers case. For instance, only 4 R-slots are required for a four carrier macrocell (1 R-slot per carrier). In case of circuit-switched operation, a lower number of R-slots is required, since access requests are performed only at call set-up. Fast-Paging-slots (FP-slots) are also required on downlink and FP-acknowledgement-slots (FP-ack-slot) on uplink. 1 FP-ack-slot is required for a single carrier macrocell and 2 for a four carrier macrocell. For data services, it may be possible to have less than 1 R-slot per frame since it is only used at call set-up and handover.

BCCH: only on one carrier per cell; not fully specified but 1 slot per frame in a macrocell and several slots per frame in a microcell are required.

DCCH: temporarily allocated to carry signalling information during handover; bidi-rectional unconstrained delay data service of 1 slot per frame; default DCCH holding times of 1 and 0.5 seconds were used for backward and forward handover procedures, respectively.

All these techniques together create a family of air interfaces (Table 2) able to provide the flexibility required for UMTS.

Cell type	Pico	Micro	short-macro	long-macro
Frame duration (ms)	5			
Slots per frame	72		18	15
Payload (symbol)	86	62 or 66		66
In-burst signalling (symbol)	10			
Training sequence (symbol)	15	33 or 29		23
Tail bits (symbol)	6	8		
Guard time (symbol)	8	12		13
Slot length (symbol)	125			120
Modulation	Binary Offset QAM			GMSK
Carrier symbol rate (kbaud)	1800		450	360
Min. carrier sep. (kHz)	1107.69 = 4x276.92		276.92	

Table 2 Family of ATDMA air interfaces. Note that all physical layer clocks can be derived from a single reference oscillator (14.4 MHz).

4. WHAT CAN BE LEARNED FROM ATDMA

The ATDMA system concept was evaluated according to the test objectives with the three different testbeds developed [5].

4.1 Technique Evaluation

Technique evaluation concentrated on the performance of individual ATDMA transport and control techniques thus giving a complete analysis of suitability of each technique for a TDMA based air-interface in UMTS.

The Transport Chain was simulated extensively, aiming at a target QoS of 10^{-6} for the constrained delay data services and a throughput of 70 % for unconstrained delay data. A large increase in magnitude of the required Quality Burst (= log10 of BER at the equaliser output), and hence a corresponding increase in C/I operating point, when the transmission delay is dropped to 30 ms compared with the 210 ms allowance for the long constrained delay data service was observed.

Adaptive Power Control showed to be well suited to be operated in combination with packet access allowing higher statistical multiplexing gains to be realised; 1.4 instead of 1.1 without power control.

Link Adaptation Computer simulations performed for the speech service showed a significant improvement of cell coverage, both, in noise- and interference limited environments. This obviously reduces the amount of base stations required (costs) to cover a certain area with a given quality constraint in noise-limited environments and allows to reduce frequency re-use distance in interference limited environments. Link Adaptation achieved the best performance in combination with packet-oriented transmission schemes. An impor-tant result was that the percentage of calls using the 2-slot mode should not be higher than 15 %, because this leads to a insufficient packet dropping rate.

When circuit-switched transmission schemes are applied the coverage gains achieved by link adaptation are partly neutralised by payload throughput losses caused by the limited number of users that can be supported by the radio resources available. This becomes in particular evident for circuit-switched data bearers with high bit rates or robust transport mode. However, link adaptation notwithstanding guarantees data transmission with high quality also under bad channel conditions - of course of the costs of capacity - where non-adaptive transmission schemes would likely fail.

Handover A very stable performance within a wide range of handover margins, mainly thanks to link control functions was obtained. DCCH signalling required by handover has a big impact on system performance, since it consumes radio resources and adds delay between the decision and the execution of the handover. Moreover, since this procedure is mobile initiated, the capacity broadcasted by the base station needs to be carefully defined to avoid high blocking at handover due to capacity congestion. However this procedure is expected to provide significant advantage in terms of cut-off probability at handover.

Packet Access evaluations (assuming a speech detector with a speech activity ratio of 45 % (mean talkspurt duration: 1.41 s, mean silence period: 1.74 s)) revealed that for a constant cell loading (i.e. mobiles are not allowed to leave their cell) a statistical multiplexing gain of 1.7 with less than 1 % packet dropping can be achieved. Allowing mobility the multiplexing gain is 1.4 - influenced by load fluctuations - which is seen as the maximum in cellular context. Admission control can dramatically reduce packet dropping but also limits overall traffic per cell. PRMA++ proved to be a generic access scheme to support required flexibility in terms of resource assignment and rapid channel assignment. On the other side it is marginally less efficient in speech-only systems. The main advantage of packet access is that while the mean occupancy of a TDMA slot is increased, the total number of slots per cell can be reduced through statistical multiplexing. Compared to circuit-switched transmission this advantage can be fully achieved after a reduction or compensation of co-channel interference through the use of advanced control techniques, either power control or link adaptation.

Furthermore operational flexibility to support different bitrate services increases significantly. The statistical multiplexing facilities are especially useful with respect to transmission of robust transport modes requiring huge amount of radio resources, because they can be easily multiplexed into periods of silence thus maintaining the same amount of payload throughput through the costs of higher slot occupancies (co-channel interference) and packet dropping pro-bability.

The DCA algorithm performed quite well with speech service using PRMA++, since channel segregation is quickly achieved. It is expected that DCA achieves a better performance than FCA in non-uniform traffic distributions, and also that DCA provides in any case a flexibility in frequency planning, that would certainly overcome any loss in capacity.

4.2 Interaction of Techniques

Within the Interaction of Techniques the project examined the performance of individual ATDMA transport and control techniques interactions.

Combination of Techniques The key parameter in any cellular mobile communication system is the *Carrier-to-Inter-ference Ratio (C/I)* in an interference limited system, or the *Signal-to-Noise Ratio (S/N)* in a noise limited system. ATDMA was designed to operate efficiently at different C/I (S/N) values. Two different approaches were followed to control C/I (S/N) ratio: an adaptive power control algorithm that sets the carrier power at the lowest possible level while still maintaining link quality and a longer term link adaptation algorithm that selects the smallest level of redundancy required to offer service integrity. As an additional aspect of interference control the exploitation of source activity, particularly for speech services, was considered; i.e. packet access technique instead of DTX. The results depicted in Fig. 3, which have been obtained from simulations with a common scenario with constant speech traffic load, simply highlight the influence of the individual ATDMA control techniques on mean service quality for circuit-switched and packet-switched transmission schemes.

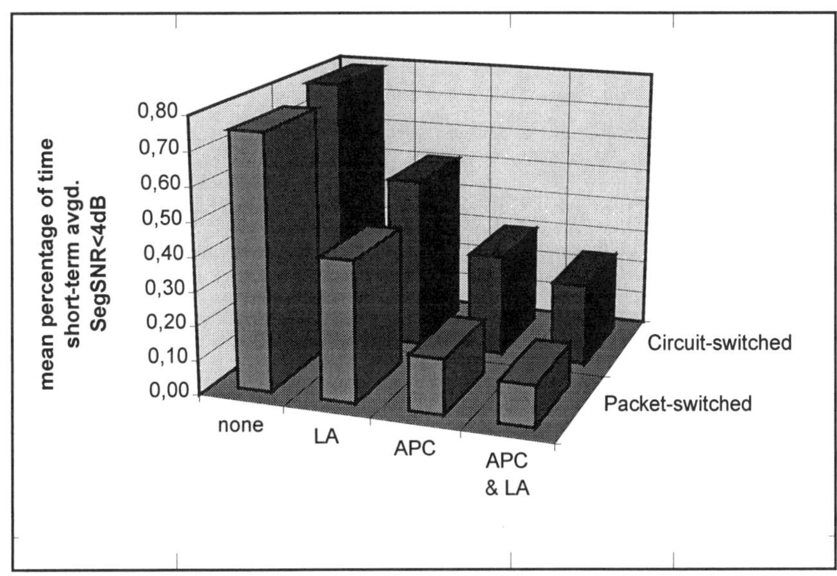

Fig. 3 Average quality improvements provided by ATDMA control technique combinations (uplink)

Enabling the combination of the ATDMA control techniques improves system performance significantly. In particular, packet switched transmissions benefit more from these techniques than circuit switched transmissions. Single quality gains provided by link adaptation and power control, respectively, do not totally complement each other, since the effects of both techniques partly neutralise each other. To summarise, all these link control techniques are successful operational in optional combinations both in packet- or circuit-switched based transmission systems, thus underlining the adaptive and flexible design concept of ATDMA.

Test Case Services The determination of the QoS parameter for each service type separately and for different service mixes resulted in the following findings:

Low transmission delay means shortened interleaving depth implying worse receiver performance due to the burst errors inherent for fading channels. This leads to a loss in system capacity due to the fact that the C/I performance of the receiver is the limiting factor for the frequency reuse and thereby capacity. To alleviate the effect of low delay frequency hopping is seen obligatory and antenna diversity optional in the ATDMA system. Especially for multi-slot transmission it is required that hopping is possible between consecutive bursts. The *target area coverage probability* has a great impact on system capacity. In a noise-limited system the advantage of gross rate link adaptation allows in principle to cut the number of required cell sites by half. In an interference-limited system the benefit of capacity depends on the quality criteria and the system structure. Furthermore the achievable system capacity can be improved significantly when relaxing the high *BER requirements*.

The results obtained by the project clearly demonstrate the difficulties in achieving high capacity simultaneously with other contradicting requirements like coverage, transmission delay and service quality. Especially the coverage requirement seems to be very tough while still keeping an acceptable quality. Therefore link adaptation to multi-slots is seen as a great advantage for the ATDMA system especially in areas where large cell areas are important. With frequency hopping between multislots high data rate services perform significantly better than low data rate services. The high BER requirements for data services, requiring typically a higher C/I than speech service, has a negative influence on the whole system performance and will imply difficulties in network planning when services with different C/I requirement should coexist in the same network.

In summary, The ATDMA system can guarantee sufficient QoS even if a high system capacity and a coverage of 99 % is demanded

4.3 System Evaluation

System Evaluation concentrated on the performance of the whole ATDMA system concept thus giving a complete analysis of suitability for an ATDMA air-interface based UMTS.

<u>System Capacity</u> The evaluation of the gains achievable by ATDMA was performed mainly in analytical studies on the basis of a *Reference System* [5]. This Reference System is a generic, circuit-switched TDMA system using the ATDMA transport chain without any of the advanced control functions but with frequency hopping. The capacity of the ATDMA system and the gain compared to 2nd generation systems and the Reference System for the speech service in macrocellular environment is shown in Fig. 4.

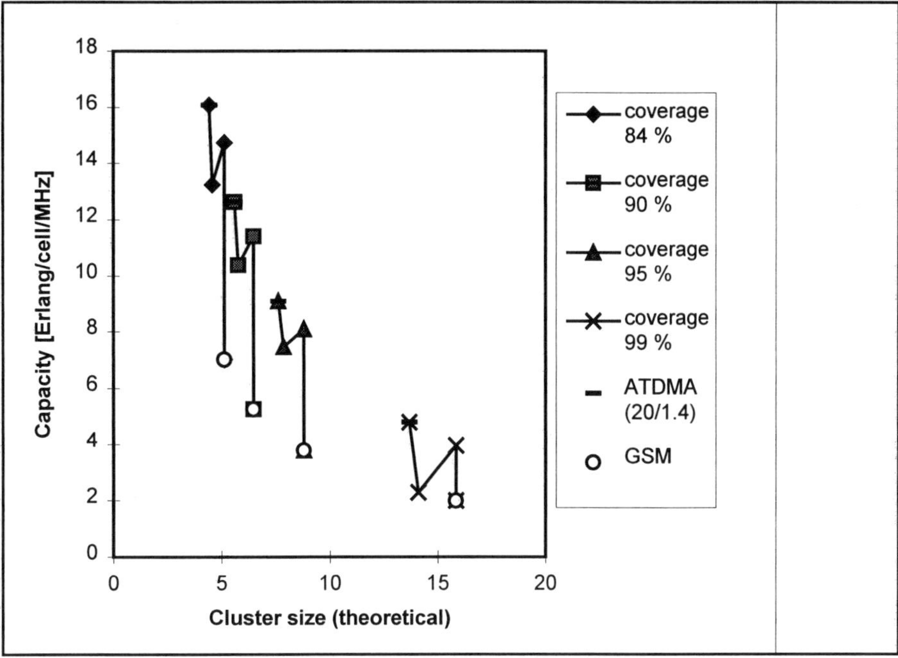

Fig. 4 Capacity of macrocellular systems versus coverage (macrocellular environment described by a pathloss slope of $\gamma = 3.6$ and a log-normal fading with standard deviation $\sigma = 6.6$ dB, blocking probability of 1 %, signalling neglected. ATDMA (20/1.4) stands for 20 % 2-slot speech transport mode usage and for packet access statistical multiplexing gain of 1.4. That gave a slot occupancy of 74 % and based on that an interference gain due to frequency hopping of about 0.5 dB could be considered (a FH gain of 2 dB has been assumed for GSM and the Reference System) The lowest point in each graph represents GSM (marked with O), the second point GSM half rate, the third point ATDMA Reference System, and the fourth and highest point of each graph represents ATDMA (20/1.4)(marked with -).

Capacity gains of 2 and more over GSM result first from ATDMA's burst and frame structure (ATDMA reference system) and second from its sophisticated control techniques, namely link adaptation and packet access (PRMA++). The (fully equipped) ATDMA system can be operated at a minimum C/I of about 5.9 dB. For speech service as minimum 8.1 dB and 3.1 dB for 1 slot and 2 slot service, respectively.

An evaluation of the potential of ATDMA in terms of capacity compared to other 3rd generation systems for the speech service and long delay data service (64 kbit/s) for a static scenario revealed the results given in Fig.5.

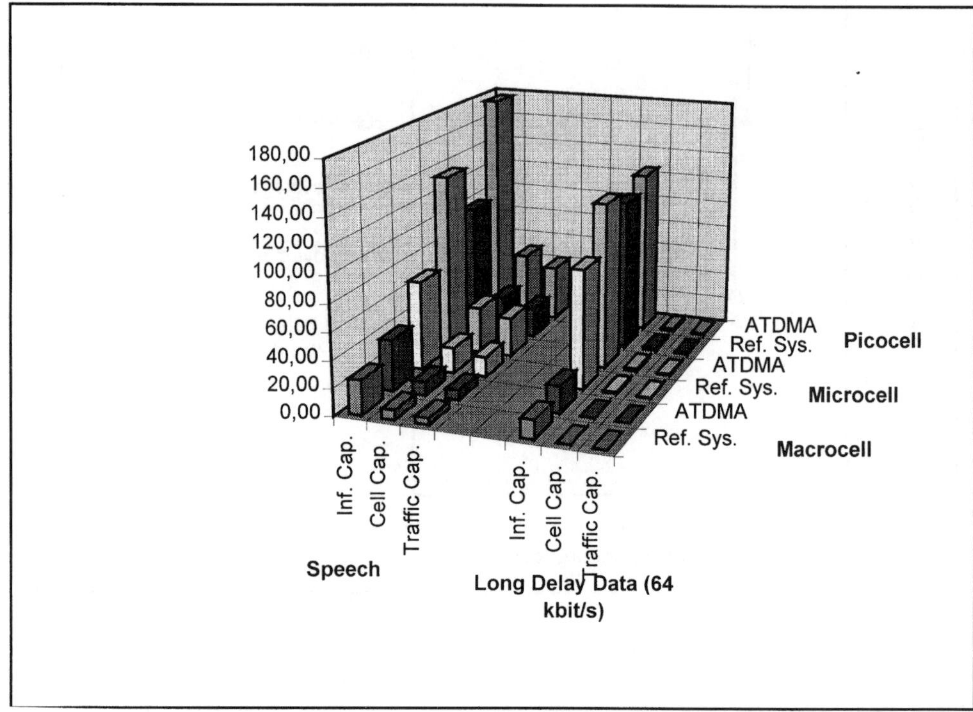

Fig. 5 ATDMA capacity figures for the speech and long delay data (64 kbit/s) services (Cell Capacity [user/cell/MHz], Information Capacity [kbit/s/cell/MHz], Traffic Capacity [Erlang/cell/MHz])

The ADTMA system performs much better in microcell and picocell scenarios, mainly because the deployment schemes offer a very good protection against intercell-interference. The advanced control functions of the ATDMA system - mainly power control and link adaptation - can then help to keep acceptable quality for all calls, thus providing high capacity gains. In particular PRMA++ offers a high multiplexing gain while keeping acceptable clipping statistics (less than 1 %).

The results obtained for the 64 kbit/s long delay data service showed that high bitrate data services with high quality targets (BER of 10^{-6}) have an heavy impact on overall system capacity. Together with UMTS' high target area coverage probability of 99 % overall capacity degraded dramatically when compared to the speech service. Such services are furthermore difficult to handle in terms of resource allocation and call admission.

Complexity Complexity in terms of signalling load required depends on the environment, the expected traffic mix, and the number of channels per base station. Assuming only speech service and taking into account both inband-signalling and signalling for control, signalling represents 59 % of the bandwidth in macrocell, 43 % in microcell and 57 % in picocell. Despite this apparent high signalling overhead the resulting gain in capacity compared to the reference system clearly demonstrates that this increase of complexity of the system is worth it.

5. ATDMA - A STEP TOWARDS UMTS

The ATDMA air-interface concept has proved that it contains the potential to offer significant gains in overall system capacity over existing TDMA systems thus representing a real enhancement compared to 2nd generation systems such as GSM. This has been reached through a flexible system concept which mainly comprises an adaptive air interface which supports different cell types, environments, services, and mobility. This adaptive approach has been realised through radio network control techniques with some self-adapting capabilities mainly designed and suited to cope with packet-switched transmission schemes. These control techniques proved their performance in extensive computer simulations and are successful operational in both packet- and circuit-switched transmission concepts, for ATDMA however packet-switched transmission is recommended.

This flexible concept can be enhanced further by e.g. macro-diversity, schemes for handling of traffic load, and hybrid access (FDMA/TDMA/CDMA, e.g. JD-CDMA). Thus this proposal is a useful approach to enhance existing TDMA systems, i.e. evolute GSM to UMTS.

Overall the ATDMA air-interface represents a promising candidate for a future UMTS, offering a wide range of services with high quality, high coverage and high capacity in all kinds of environment thus coping with the needs of the future UMTS user.

ACKNOWLEDGMENT

The author acknowledges the contributions of the colleagues from the RACE II ATDMA consortium and the support of the Commission of the European Union. Views expressed in this contribution are the view of the author and not necessarily the views of the RACE 2084 project as a whole.

REFERENCES

[1] Cygan, D.; David, F,; Eul H.-J.; Hofmann, J.; Metzner, N.; Mohr, W.; Nottensteiner, H.: RACE-II Advanced TDMA Mobile Access Project - An Approach for UMTS. International Zürich Seminar on Digital Communications, Zurich, Switzerland, 08 - 11 March 1994, Proceedings, p. 428 - 439

[2] Mourot, Ch.; Streeton, M.; Urie, A.: An Advanced TDMA Mobile Access System for UMTS. IEEE Personal Communications Magazine, Vol. 2 No. 1, February 1995

[3] Grillo, D.; Metzner, N.; Murray E.D.: Testbeds for Assessing the Performance of a TDMA-Based Radio Access Design for UMTS. IEEE Personal Communications Magazine, Vol. 2 No. 2, April 1995

[4] DeVile, J.: A Reservation Multiple Access Scheme for an Adaptive TDMA Air-Interface. 4th WINLAB Workshop on 3rd Generation Wireless Information Networks, New Jersey, USA, October 1993, Proceedings, p. 217 - 225

[5] David, K.; Blanc, P.; Irvine, J.; Kassing, T.; López-Carrillo, M.I.; Maddalena, R.; Mohr, W.; Ranta, P.: First Results of ATDMA's Test Campaign. RACE Mobile Telecommunication Summit, Cascais, Portugal, 22 - 24 November 1995, p. 468 – 472

Chapter 2

UMTS Data Services

Tero Ojanperä
Nokia Research Center

1. INTRODUCTION

The tremendous success of GSM has created a need for evolution of the services towards higher bit rates. Universal Telecommunication System (UMTS), a third generation wireless system, targeting towards maximum peak bit rate of 2 Mbit/s, is a natural continuation for the development of GSM High Speed Circuit Switched Data Service (HSCSD) with maximum bit rate of 115.2 kbit/s, and General Packet Radio Service (GPRS) providing maximum bit rates between 89.6 and 182.4 kbit/s depending on the applied channel coding.

ETSI (European Telecommunication Standardization Institute) started the standardization of UMTS in 1991. Within ETSI, Technical Committee SMG (Special Mobile Group) is responsible for UMTS standardization. Sub Technical Committee (STC) SMG2 is in charge of the UMTS radio access system standardization and SMG3 of the development of possible new UMTS/GSM core network standards. In addition, other STCs will be involved in timely manner as the detailed standardization proceeds. First years of the UMTS standardization were devoted to technical studies which have been condensed into a number of ETSI Technical reports [1]-[13]. Based on these reports the development of UMTS standards has started [14]-[15]. Furthermore, the UMTS Terrestrial Radio Access Concept (UTRA) definition process was initiated in a workshop held in December 1996 by introduction of number of candidate radio access schemes. According to current plans of SMG, decision of the UTRA concept will be made in the end of 1997 and main technical parameters will be frozen by mid 1998.

ITU (International Telecommunication Union) is developing common vision for third generation wireless system which is called IMT-2000, International Mobile Telecommunications (previously it was called FPLMTS, Future Public Land Mobile Telecommunication Systems). ITU-R TG8/1 has initiated the selection process of IMT-2000 radio transmission technologies.

World-wide, the process of standardising third generation radio access has proceeded fastest in Japan. The Japanese standards body ARIB (Association of Radio Industries and Business) has selected wideband CDMA (W-CDMA) to be the Japanese candidate scheme for the IMT-2000 air interface. Furthermore, the biggest Japanese operator NTT DoCoMo has ordered prototypes for a W-CDMA trial system.

In USA, TIA (Telecommunications Industry Association) committee TR45.3 is standardising IS-136 evolution towards IMT-2000. Proposals include TDMA and OFDM based approaches. TIA TR45.5 is standardising wideband CDMA based on evolution from IS-95.

This chapter presents issues related to introduction of UMTS data services. In 2, world wide UMTS/IMT-2000 spectrum allocations and impact of spectrum licensing on UMTS are discussed. Section 3 assesses the potential of GSM towards UMTS and Section 4 explains the GRAN (Generic Radio Access Network) concept. UMTS bearer service requirements are discussed in Section 5. An overview of UMTS multiple access alternatives is presented in Section 6. These inlcude wideband CDMA, TDMA, OFDM and hybrid CDMA/TDMA. It mainly concentrates into European and Japanese proposals since both Europe and Japan have adopted GSM as a basis for third generation core network. Furthermore, time division duplex (TDD) aspects and possible integration of wireless broadband networks with GSM are discussed. Impact of market and operator scenarios, services and regulatory aspect on transition into UMTS are presented in Section 7. In Section 8 conclusions are given.

2. SPECTRUM ALLOCATION AND LICENSING FOR UMTS

In order to create a secure environment for the investments into third generation wireless systems, a regulatory framework catering for spectrum allocation and licensing conditions should be developed. The UMTS Forum was created in 1996 in order to accelerate the process of defining the necessary standards, policy actions and industrial co-operation for UMTS [16].

Currently, UMTS Forum is developing a report on regulatory frame work for UMTS [16]. One of the most important aspects is the definition of spectrum needs for UMTS. Figure 1 presents the WARC'92 (World Administrative Radio Conference) allocation of UMTS/IMT-2000 spectrum and how it is used in USA, Europe and Japan. As can be seen, European and Japanese third generation frequency allocations are very similar. In contrast, the IMT-2000/FPLMTS spectrum in USA is already used for PCS (Personal Communication Systems). Thus, the technology choices are limited due to a lack of spectrum. In Europe, ERC (European Radio Committee) has decided that at least 2 times 30 MHz of the identified IMT-2000 spectrum should be allocated for UMTS by year 2002 [17].

The licensing policy of UMTS spectrum might also impact the technology choices. A GSM operator without UMTS spectrum needs most likely a different solution to provide third generation services within the limited GSM spectrum than an operator with both GSM and UMTS spectrum licenses.

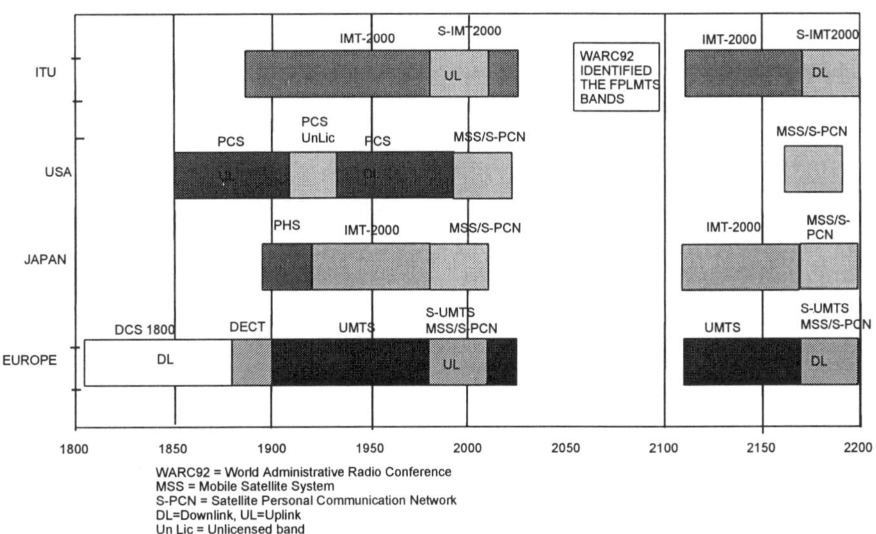

Figure 1. World-wide spectrum allocation

3. GSM - THE STEPPING STONE INTO UMTS

GSM radio capabilities are evolving in two dimensions - radio environments and service bit rates. Originally developed for macro cellular vehicular environment, GSM has evolved into a wireless system spanning all radio environments from indoor into satellite. The GSM evolution towards high data rate services can be described as waves. The first wave was the commercial introduction of GSM 9.6 kbit/s data service in 1994. The second wave will be the introduction of GSM HSCSD with a maximum data rate of 115.2 kbit/s and GPRS with a maximum data rate of 182.4 kbit/s in 1998-1999. Finally, the third wave will be the introduction of UMTS wideband services providing wide area bit rates up to 384 kbit/s and local coverage up to 2 Mbit/s by the turn of the century.

Table 1. GSM compliance for the UMTS targets .

UMTS target	GSM comply?
Small affordable handportables	YES
Deep penetration (beyond 50 %)	YES, over 30 % today
Any-where any time (indoor, office)	YES with pico cells and GSM office concept
Anywhere - satellite mobile interworking	YES, dual mode
Hot spot capacity	YES, by cell hierarchies,
Wireline voice quality	YES, EFR codec
Global roaming	YES, SIM, MAP
IN services	YES, CAMEL
Multimedia, entertainment, non-voice	YES, TCP/IP transparency, GPRS, HSCSD
Flexibility to mix different bearer types (non-real time and real time)	NOT YET, UMTS
High bit rate services (over 200 kbit/s)	NOT YET, UMTS

In contrast to speech services, the market requirements and user needs for data services are not well understood, yet. Increased possibilities to offer higher rate services with different quality of service create, beside opportunities, also challenges to be able to market them to the end users. The above described three waves, starting from the basic GSM data proceeding via GPRS and HSCSD into UMTS, facilitate proper learning to market and use the new data services with step wise investments utilising the capabilities of the existing GSM networks.

The enormous growth of wireless subscriber base and introduction of large number of new services calls for a robust and flexible network architecture. The well standardised, totally open multivendor GSM system has proven to be very attractive for the operators; the network extensions will be easy and not so costly as from buying from a single vendor. For

UMTS this will be even more important. Thus, the well proven GSM network principles with enhanced capabilities will be adopted for UMTS.

It is interesting to view the third generation requirements in the light of the latest GSM developments. As Table 1 shows, all other targets except the flexibility to mix real-time and non-real time within same connection and the high bit rate services beyond 200 kbit/s are met by GSM. Thus, adding a new radio interface providing the high bit rate services with increased flexibility on top of other GSM capabilities is the easiest and most profitable way into the third generation.

4. GENERIC RADIO ACCESS (GRAN) CONCEPT

UMTS will be based on the concept of Generic Radio Access Network (GRAN) capable of connecting into several core networks. The GRAN concept, depicted in Figure 2, can be connected into GSM/UMTS, N-ISDN, B-ISDN and packet data networks. Furthermore, the GSM BSS can be connected into the GSM/UMTS core network. This relationship between GSM and UMTS is important, especially in the beginning of the UMTS life time, in order to facilitate a spectrum efficient utilisation of GSM900, GSM1800, GSM1900 and UMTS frequency bands and seamless coverage using multimode GSM/UMTS terminals. The evolved GSM core network facilitates also service portability and full utilisation of the existing investments and customer base.

Similar to GSM air interface (Um), the UMTS air interface will be an open well specified standard (Uu interface). Furthermore, the Iu interface specifies the interface between GRAN and core networks and Cu interface terminal equipment and UMTS subscriber identity module (USIM).

5. BEARER SERVICE REQUIREMENTS FOR UMTS

During the recent years, two hottest topics in telecommunications have been the growth booms of Internet and cellular systems. Integration of these two, together with new wideband transfer capabilities, will form the cornerstones of UMTS.

Internet will provide multimedia services: audio, data and video together. Thus, the transmission requirements such as bit rates, bit error rate (BER) and delay are highly variable. The services delivered through UMTS radio access system can be classified as real-time and non real-time. Generic bearer capabilities for UMTS are listed in Table 2. It should be noted that the exact values are currently under discussion in the standardization bodies.

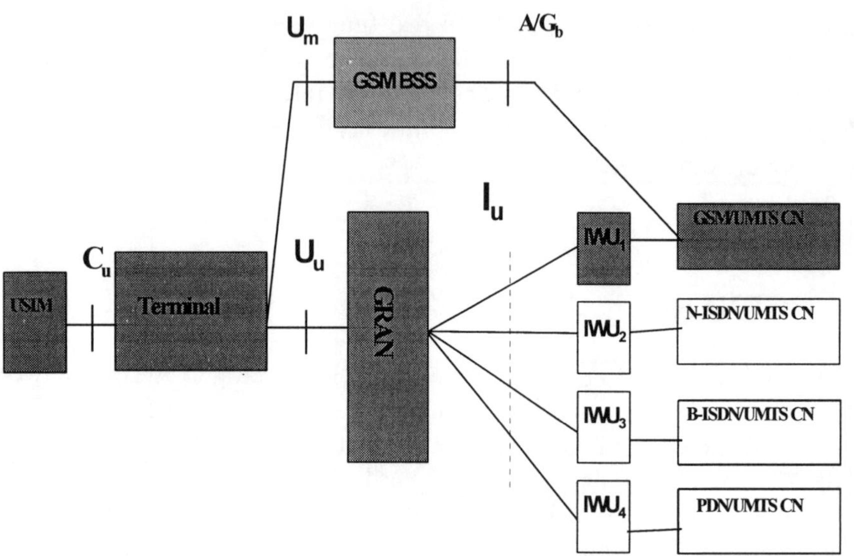

Figure 2. GRAN concept

In addition to the above basic bearer requirements, number of requirements have been set how to use the UMTS bearers. The UMTS radio access system should be able to negotiate the bearer service attributes such as bearer type, bit rate, delay, BER, up/down link symmetry and protection including none or unequal protection. It should be possible to associate several bearers with a call and bearers can be added and released during a call. UMTS should also able to adapt the bearer service bit rate, quality and radio link parameters depending on the link quality, traffic, network load, and radio conditions within the negotiated service limits.

Table 2. UMTS Bearer Service Requirements.

	Real Time/Constant Delay		Non Real Time/Variable Delay	
Operating environment	Peak Bit Rate	BER / Max Transfer Delay	Peak Bit Rate (note 3)	BER / Max Transfer Delay
Rural outdoor (terminal speed up to 250 km/h)	384 kbit/s granularity appr. 10 kbit/s	BER 10^{-3} - 10^{-7} delay 20 - 300 ms	384 kbit/s	BER = 10-5 to 10-8 Max Transfer Delay 150 ms or more
Urban/ Suburban outdoor (Terminal speed up to 150 km/h)	512 kbit/s granularity appr. 10 kbit/s	BER 10^{-3} - 10^{-7} delay 20 - 300 ms	at least 384 kbit/s (preferably 512 kbit/s)	BER = 10^{-5} to 10^{-8} Max Transfer Delay 150 ms or more
Indoor/ Low range outdoor (Terminal speed up to 10 km/h)	2 Mbit/s granularity appr. 100 kbit/s	BER 10^{-3} - 10^{-7} delay 20 - 300 ms	2 Mbit/s	BER = 10^{-5} to 10^{-8} Max Transfer Delay 150 ms or more

6. MULTIPLE ACCESS ALTERNATIVES

The UMTS Terrestrial Radio Access (UTRA) will be defined in a comparative evaluation process based on the UMTS radio requirements listed in [7]. In the first phase of this process, the different candidate schemes have been grouped into five concept groups: wideband CDMA, TDMA, hybrid CDMA/TDMA, OFDM (orthogonal frequency division multiplexing) and ODMA (opportunity driven multiple access). In this section the multiple access research activities related to these schemes and an overview of wideband CDMA, TDMA, OFDM, and hybrid CDMA/TDMA are presented.

6.1 Multiple Access Research

In Europe, RACE I (Research of Advanced Communication Technologies in Europe) program, launched in 1988 and lasted until June 1992, started the third generation research activities. During 1992 and 1995, in the RACE II program, CODIT (Code Division Multiple Testbed) and ATDMA (advanced TDMA) projects developed air interface specifications and testbeds for UMTS radio access [18]-[21]. In addition, several industry projects have developed air interface concepts and trial systems for UMTS [22]-[26]. European Research Program ACTS (Advanced Communication Technologies and Services) was launched at the end of 1995 to support collaborative mobile research and development. Within ACTS the project FRAMES (Future Radio Wideband Multiple Access System) has been set up with an objective to define a proposal for UMTS radio access system. The first target of FRAMES was to investigate hybrid multiple access technologies and based on thorough evaluation of several candidate schemes select the best combination as a basis for further detailed development of UMTS radio access system. The comparison results have been presented in [27]-[29]. Based on this comprehensive evaluation, a harmonized multiple access platform FRAMES Multiple Access (FMA) has been designed consisting of two modes: FMA1, a wideband TDMA with and without spreading and FMA2, a wideband CDMA [30]-[32]. FMA2 has been submitted both to ETSI and ARIB.

FPLMTS Study Committee in ARIB was established in April 1993 to co-ordinate the Japanese research and development activities for IMT-2000/FPLMTS system. In October 1994, it established Radio Transmission Special Group for radio transmission technical studies and production of draft specification for FPLMTS [33]. The Special Group consists of two Ad Hoc groups: CDMA and TDMA [34]. Originally, 13 different wideband CDMA/FDD (Frequency Division Duplex) radio interfaces were presented to FPLMTS Study Committee. In the beginning of 1995, they were merged into three CDMA/FDD proposals, Core A, B and C, and into one TDD proposal. In the end of 1996, the four schemes were further combined into a one single proposal where the main parameters are from the Core-A [35]-[39]. Three of the original wideband CDMA schemes closely resembling the Core-A have been submitted into ETSI as well. In TDMA group wideband TDMA and OFDM schemes have been developed but they are not continued for further standardization. However, the OFDM scheme called BDMA (Band Division Multiple Access) has been submitted to ETSI.

6.2 TDMA

Table 3 presents the main characteristics of two TDMA based air interfaces proposed for UMTS: FMA1 without spreading (wideband TDMA) [31] and a GSM compatible ATDMA scheme [40]. Furthermore, we cover the proposed GSM evolution using higher level modulation (EDGE, enhanced data services for GSM evolution) also called GSM384 reflecting its capabilities to deliver data rates up to 384 kbit/s and beyond [41]. FMA1, wideband TDMA is intended for high bit rates and especially packet traffic. ATDMA concept was started originally from a clean table without any backward compatibility in mind. However, when the idea of GSM backward compatibility emerged as one of important criteria for the UMTS air interface, ATDMA basic concept was modified to be GSM backward compatible. The EDGE concept improves the GSM data rates with higher level modulation such as 16-QAM within the current 200 kHz carrier bandwidth. By employing the high level modulation scheme and a higher symbol rate, EDGE achieves a gross rate approximately three times higher than GSM. By combining several time slots a user rate of 384 kbps can be reached

All TDMA based proposals have GSM compatible frame structure with 4.615 ms frame length. For very high data rate services a wider bandwidth is used. In this case the number of slots is large to accommodate the low bit rate services as well. Multislot and multi level modulation are used to increase the bit rate. Frequency hopping provides frequency diversity as well as interference diversity through interleaving and coding.

In addition to Frequency Division Duplex (FDD) mode, FMA1 and ATDMA have Time Division Duplex (TDD) mode.

More advanced radio resource algorithms such as DCA (dynamic channel allocation) and link adaptation will facilitate adaptation into different carrier to interference ratios required by different services. One additional performance improvement method is Joint Detection (JD) for reduction of other cell interference [42]. Additional benefit of Joint Detection is that spectrum efficiency is not sensitive for reuse factor which reduces the need for network planning.

Based on the above three TDMA proposals we can summarise a possible TDMA based evolution track from GSM into UMTS into three phases

- GSM phase 2+, HSCSD, GPRS
- EDGE/GSM384
- wideband TDMA

Table 3. TDMA based air interface schemes

	FMA1 - wideband TDMA		ATDMA (GSM compatible)				Enhanced GSM
	data burst /regular burst	micro cell sub burst	Long macro (GSM)	Short macro	Micro	Pico	
Carrier Spacing	1.6 MHz	1.6 MHz	200 kHz	200 kHz	800 kHz	1600 kHz	200 kHz
Carrier Bit Rate	2.6 Mbit/s 5.2 Mbit/s	2.6 Mbit/s 5.2 Mbit/s	271 kbit/s	325 kbit/s	1.3 Mbit/s	2.6 Mbit/s	325 kbit/s 770 kbit/s
Duplex mode	FDD/ TDD	FDD/ TDD	FDD/ TDD	FDD/ TDD	FDD/ TDD	FDD/ TDD	FDD
Modulation	B-O-QAM Q-O-QAM	B-O-QAM Q-O-QAM	GSMK	B-O-QAM Q-O-QAM	B-O-QAM Q-O-QAM	B-O-QAM Q-O-QAM	B-O-QAM TC16-QAM
Frame Length	4.615 ms	4.615 ms	4.615 ms	4.615 ms	4.615 ms	4.615 ms	4.615 ms
Number of Slots Per Frame	16/64	64	8	10	40	80	8
Payload (bits)	684/ 122	144	114+2	114+2	114+2	114+2	148
Training Sequence (symbols)	49 (18.8 μs)	27 (10.3 μs)	26	19	19	19	26
Tail Bits	3	3	6	6	6	6	6
Guard Time (symbols/ us)	11 (4.2us)/ 10.5 (4us)	10.5 (4μs)	8.25 (30.4μs)	9 (26.7μs)	9 (6.7μs)	9 (5 μs)	7.5
Max. Tolerable Delay Spread	7 μs	2.7 μs	over 20 μs	20 μs	5 μs	1 μs	-

Furthermore, advanced radio resource management algorithms and performance increase methods such as joint detection can be introduced to combine EDGE and wideband TDMA into full scale TDMA based UMTS air interface consisting of both narrowband and wideband carriers.

The use of TDMA based UMTS solution depends then on the operator and market scenarios. An existing GSM or GSM1800 operators without UMTS frequency license as well as a PCS1900 operator would use EDGE with joint detection (the narrowband UMTS carrier) to obtain better spectrum efficiency and to introduce higher bit rate services. An existing GSM operator with UMTS spectrum license could use only wideband TDMA to introduce high bit rate services. A new UMTS operator would use TDMA based UMTS both with narrowband and wideband carriers to provide full set of UMTS services for wide area and ho-spots coverage. Moreover, any of the above mentioned operators could use the TDD mode of wideband TDMA to introduce services in the UMTS TDD frequency bands for home or office use.

EDGE and wideband TDMA have been proposed also in TR45.3 for IS-136. Thus, we could see a common high speed physical layer for IS-136 and GSM. This would be beneficial from the economies of scale perspective. Both systems would, however, retain their own higher layer protocols.

6.3 Hybrid CDMA/TDMA

FRAMES FMA1 with spreading is based on hybrid CDMA/TDMA concept, also called joint detection [44]-[46]. The role of CDMA is to multiplex the different channels within a timeslot. Furthermore, the CDMA component facilitates use of RAKE receiver and more advanced receiver techniques such as joint detection. In a multipath channel with a large system load, joint detection is actually mandatory in order to remove the intracell interference.

Original parameters of joint detection CDMA, frame length of 6 ms and 12 slots per frame [44], have been aligned with FMA1 without spreading (see 6.2) to be backward compatible into GSM, i.e., 4.615 ms and 8 slots per frame, respectively. The main difference of FMA1 with spreading and FMA1 without spreading (wideband TDMA) is, of course, spreading. Short orthogonal spreading codes of length 16 are used. Spreading modulation method is GMSK and data modulation QPSK.

FMA1 with spreading uses either multicode or multislot to increase the user bit rate. Within each slot maximum 16 codes can be pooled together for one user. However, typically only 10 of the 16 codes can be used due to interference limitation [43].

Code division in FMA1 with spreading can be viewed as a multiplexing method facilitating transmission of low and high bit rate services within the same RF bandwidth. The drawback of this approach is spectrum regrowth due to multicode transmission and non-linear power amplifiers [43].

The training sequence length of Mode 1 is adapted to the expected channel impulse response length. There are bursts with two different training sequence lengths. The longer training sequence is suited for estimating the 8 different uplink channel impulse responses of 8 users within the same time slot with a time dispersion of up to about 15 μs. If the number of users is reduced, the tolerable time dispersion is increased about proportionally. The shorter training sequence is suited for estimating the 8 uplink channel impulse responses with a time dispersion of up to about 5.5 μs. Furthermore, it is suited for estimating the downlink channel impulse response with a time dispersion of up to about 25 μs, independent of the number of active users and the uplink channel impulse response with this same time dispersion in case all bursts within a slot are allocated to one and the same user [43].

6.4 Wideband CDMA

In the UTRA concept group evaluation process, the FMA2 proposal was considered together with the wideband CDMA schemes from Japan. At the same time, International Co-ordination Group (ICG) in ARIB had discussions on the harmonisation of different CDMA proposals. These two efforts led to the harmonisation of the parameters for ETSI and ARIB wideband CDMA schemes. The main parameters of the current scheme are based in the uplink on the FMA2 scheme and in the downlink on the ARIB wideband CDMA. Also, contributions from other proposals and parties have been incorporated to further enhance the concept.

Wideband CDMA has several attractive properties over second generation narrowband CDMA. In a radio channel the original signal is reflected from obstacles such as houses, hills etc. and the receiver gets several copies of the signal. Wideband CDMA can resolve these signals and is able to combine them to improve the performance. Furthermore, wideband CDMA is able to average interference from other users more effectively. This is especially important for high rate data users.

Wideband CDMA signal structure is well suited for multiplexing of different services with different quality of service requirements. WCDMA can vary the bit rate and service parameters on frame-by-frame basis. The multi rate concept is based on variable spreading for low and medium bit rates and multicode for the highest bit rates.

In addition to the fast power control in uplink, WCDMA has fast power control also in the downlink. This will improve performance against fading channel.

Table 4. Key features of WCDMA proposals

Channel Bandwidth	5, 10, 20 MHz
Downlink RF channel structure	Direct spread
Chip rate	4.096/8.192/16.384 Mcps
Frame length	10 ms / 20 ms (optional)
Spreading modulation	Complex quadrature spreading
Coherent detection	User dedicated time multiplexed pilot
Multirate	Variable spreading and multicode Time multiplexed simultaneous services
Spreading factors	4-256 (4.096 Mcps)
Power Control	Open and fast closed loop (1.6 kHz)
Spreading	Variable length orthogonal sequences for channel separation, Gold sequences for cell and user separation

Coherent detection in the uplink increases performance by 3 dB compared to non-coherent detection. Coherent detection is performed with help of user dedicated pilot symbols. Moreover, user dedicated pilot symbols in the downlink ease the implementation of adaptive antennas.

WCDMA has asynchronous cell sites. An asynchronous network does not require any external timing reference such as a GPS (Global Position System). GPS would increase costs since it requires a separate receiver and, for indoor solutions, an outdoor antenna installation to catch the reference signal. Thus, wideband CDMA with an asynchronous network is a cheaper and more reliable solution compared to synchronous network.

In a high capacity network there will be several carrier frequencies and hierarchical cell structures. WCDMA has seamless interfrequency handover for handover between two different carrier frequencies. Seamless interfrequency handover has two implementation alternatives: compressed mode/slotted mode and dual receiver [43].

WCDMA supports several carrier bandwidths: 5, 10, and 20 MHz. However, the main bandwidth alternative is 5 MHz providing bit rates up to 384 with appropriate coding for BER 10-3 and 10-6.

6.5 OFDM (Orthogonal Frequency Division Multiplexing) based schemes

Introduction of OFDM into cellular world has been driven by two main benefits:

- flexibility: each transceiver has access to all subcarriers within a cell layer
- easy equalization: OFDM symbols are longer than the maximum delay spread resulting into flat fading channel which can be easily equalised.

Also the introduction of Digital Audio Broadcasting (DAB) based on OFDM and research of OFDM for HIPERLAN type II (High Performance Local Area Network) and Wireless ATM (Asynchronous Transfer Mode) have increased the interested towards OFDM [47].

Main drawback of OFDM is the high peak to average power. This is especially severe for the mobile station and for long range applications. Different encoding techniques have been investigated to overcome this problem. Furthermore, the possibility to access to all resources within the system bandwidth results into an equally complex receiver for all services, regardless of the bit rate. Of course, a partial FFT for only one OFDM block is possible for low bit rate services, but this would require a RF synthesiser for frequency hopping.

For UMTS and FPLMTS/IMT-2000 two OFDM air interface concepts have been presented: BDMA [33] and OFDM by Telia (Table 5). BDMA concept has actually been proposed both in ETSI and in ARIB FPLMTS Study Committee. More information from Telia's OFDM concept can be found from [48]-[51] Main difference between the Telia concept and BDMA is the detection method which impacts the overall design of the systems. Telia OFDM uses coherent detection and BDMA differentially coherent detection. Telia OFDM employs coherent detection due to two reasons: performance gain and opportunity to use arbitrary signal constellations such as 16-QAM [26].

Table 5. Main features of OFDM proposals.

	Telia OFDM	BDMA
Bandwidth	5 MHz	5 MHz
number of subcarriers	1024	800
OFDM subcarrier bandwidth	5 kHz	6.25 kHz
OFDM symbol length	200 μs	200 μs
OFDM block size	25 carriers × 3 symbols	24 carriers × 1 symbol
Frame length	15 ms	5 ms, divided into 4 subframes
Detection	coherent	differentially coherent

6.6 TDD mode

Main discussion about UMTS air interface has been around technologies for frequency division duplex (FDD). However, there are several reasons why also time division duplex (TDD) could be used. First of all, there will be most likely a dedicated frequency bands for TDD within the identified UMTS frequency bands. Furthermore, FDD requires exclusive paired bands and spectrum for such systems is therefore hard to find. On the other hand, TDD can make use of individual bands which do not need to be mirrored for the return path, and hence spectrum is more easily identified. With a proper design including powerful FEC TDD can be used even in outdoor cells. It has been argued that the TDD guard interval would result to excessive overhead in large cells. However, in a cell with a range of three kilometres we would need a 20 μs guard time to prevent transmission and reception time slots to overlap, i.e., an overhead of approximately 4 % assuming frame length of 4.615 ms. When the propagation delay exceeds the guard period, soft degradation of performance occurs. Thus, UMTS TDD mode need not to be restricted into unlicensed indoor solutions but perhaps even most short-range UMTS should be TDD, even that used by the traditional cellular operators, for example in high-capacity microcells. Second reason for using TDD is the flexibility in radio resource allocation, i.e., bandwidth can be allocated by changing the number of time slots for up and downlink.

The asymmetric allocation of radio resources leads into two interference scenario that will impact the overall spectrum efficiency of TDD scheme:

- asymmetric usage of TDD slots will impact the radio resource in neighbouring cells
- asymmetric usage of TDD slots will lead into blocking of slots in adjacent carriers within own cell

Figure 3 depicts the first scenario. Mobile station 2 (MS2) is transmitting at full power at the cell border. Since the mobile station 1 (MS1) has different asymmetric slot allocation than the mobile station 2 its' downlink slots received at the sensitivity limit are interfered by the mobile station 1 causing blocking. On the other hand, since the base station 1 (BS1) can have much higher EIRP (effective isotropically radiated power) power than mobile station 2, it will interfere the base station 2 (BS2) receiving mobile station 2. Hence, the channel allocation algorithms need to avoid this kind of situation.

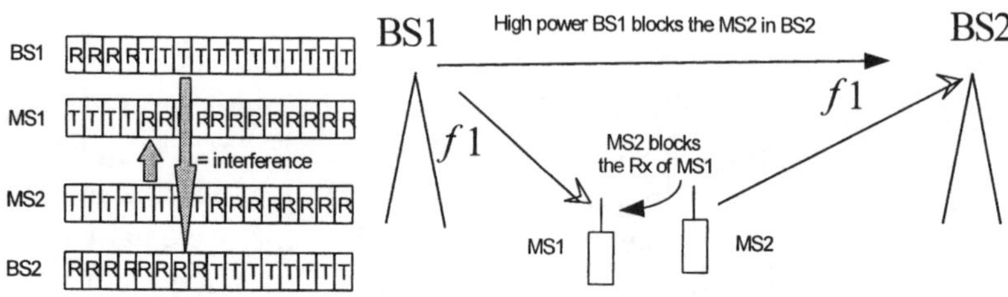

Figure 3. TDD interference scenario

In the second scenario, two mobiles would be connected into same cell but using different frequencies. The base station is receiving mobile station 1 on the frequency f1 using the same time slot it uses on the frequency f2 to transmit into mobile station 2. The transmission will block the reception due to irreducible noise floor of the transmitter regardless of the frequency separation between f1 and f2.

Third scenario, where the above described blocking effect exists, is an FDD system where if at any moment traffic in a cell is unbalanced between the up-link and down-link then the spare capacity in the low-traffic direction might momentarily be used for a two-way operation, i.e., TDD.

DECT (Digital European Cordless Telephone) is a second generation TDD system with carrier spacing of 1728 kHz and carrier bit rate of 1152 kbit/s. DECT frame length is 10 ms and each frame is divided into 24 slots. DECT will provide bit rates up 512 kbit/s half duplex and 256 kbit/s full duplex. The fundamental difference between DECT and UMTS are the bit rate capabilities, operating point and channel coding. Since the UMTS TDD mode has powerful channel coding, the required C/I and hence also reuse factor is smaller compared to DECT. Therefore, the assumption of similar up and downlink interference situation does not hold anymore, and the DECT Dynamic Channel Selection (DCS) is not suitable for UMTS. Also the performance of DECT in high delay spread environments is not very good which limits the outdoor cell range.

Both TDMA and CDMA based schemes have been proposed for TDD. Most of the TDD aspects are common to TDMA and CDMA based air interfaces. However, in CDMA based TDD systems we need to change symmetry of all codes within one slot in order to prevent interference situation where high power transmitter would block another receiver. Thus,

TDMA based solutions have higher flexibility. In FRAMES Multiple Access only FMA1 has a TDD option which can be used both with wideband TDMA without spreading and with spreading [30]. However, with spreading option the above mentioned drawback of code allocation exists. CDMA/TDD has been proposed in [33].

6.7 Wireless Broadband Networks

Wireless Broadband Networks will provide user bit rates up to 20 Mbit/s. Wireless ATM (Asynchronous Transfer Mode) is the most promising technology to implement the Wireless Broadband Networks. The basic idea of wireless ATM (WATM) is to extend the communications capabilities of wired ATM such as provision of different Quality of Service classes into mobile users [47].

ETSI is currently developing high speed WATM radio standard, HIPERLAN Type 2, for the 5 GHz frequency band. The cell range of WATM system would be in the order of the 20 to 50 meters. There is 200 MHz unlicensed spectrum available both in USA and Europe. This facilitates simple air interface design because hardly any attention needs to be paid into spectrum efficiency as in the cellular systems using licensed frequency bands. Furthermore, the tariffing structure for high speed data services could be different from the cellular since there is no price for the spectrum and allocating spectrum for data services does not decrease the revenues from other services such as speech.

The 20 Mbit/s data rate and 200 MHz unlicensed spectrum make the integration of WATM radio access with GSM and UMTS an attractive opportunity to extend the current cellular business into a new market. The GSM/UMTS would offer wide area mobility and coverage while the WATM would offer hot spot broadband data services.

7. TRANSITION INTO UMTS

In this section we study the impact of services, market and operator scenarios as well as regulatory aspects such as spectrum allocation into the introduction of UMTS high bit rate services. Previously, the various aspects of the evolution towards UMTS including techno-economics, marketing, regulation and licensing have been considered in [4].

There exists two main market scenarios for the introduction of UMTS data services. In the first scenario demand for high bit rate services is secure and thus the investments can be financed by incomes. In the second scenario the data demand is unsecured and operator need to catch the data users with a minimum investment into equipment, frequency spectrum etc. This might impact the air interface technology choice since some technologies require initially more spectrum than others for the same service set.

Operator has to decide what bit rates are satisfactory for the applications and users; is 144 kbit/s enough for wide area coverage or should it be 384 kbit/s. And what impact does the bit rate has on the user's behaviour; is there a breakpoint in the user bit rate where users start to use a certain data service more because access becomes more convenient. Furthermore, the service type will impact the deployment. A unconstrained data transfer will tolerate large variation in the bit rate during connection facilitating a network deployment where high data rates are only available near the cell site. On the other hand, good quality video connection requires the same data throughput everywhere.

There exists three scenarios for the UMTS deployment

- UMTS introduced into UMTS frequency band
- UMTS introduced into GSM frequency bands
- GSM introduced into UMTS frequency bands

An existing cellular operator will most likely start UMTS with local coverage relying on GSM for low bit rate wide area coverage. Therefore, it is wise to ease the dual mode terminal design by properly selected air interface parameters. In the longer term the replacement of GSM by UMTS depends on the gained advantage such as higher bit rates or increased spectrum efficiency. Furthermore, licensing and availability of spectrum also impact the decision. In case the available bands of an operator are congested, the higher spectrum efficiency of UMTS allows the operator to pack existing users into less spectrum and to create room for higher rate services. However, the introduction of a wideband carrier into congested frequency bands is fairly difficult. First, the operator needs preload the network with dual mode terminals with new wideband capabilities. Next, he needs to release the spectrum from old carriers and switch the new carrier on almost simultaneously to avoid interruption in service.

Green field UMTS operator in country with an existing cellular network needs also a dual mode terminals. Only if the data market emerges very fast and justifies investment into a new nation-wide network operator could rely on single mode terminals.

Transition into UMTS depends on technical possibilities such as increased spectrum efficiency and data rates but most of all on market needs and regulatory conditions.

8. CONCLUSIONS

UMTS data services will be introduced in the turn of the century. The separation of core network and radio access network facilitates independent evolution of both parts. Any radio access technology can be interfaced with the GSM/UMTS core network, the most dominating core network globally.

In Europe UMTS air interface will be selected in the beginning of 1998. The target has been and still is to select only one air interface. Therefore, from the global perspective, a natural alternative would the asynchronous wideband CDMA with 5 MHz bandwidth offering wide area data rates up to 384 kbit/s. Alternative technologies are wideband TDMA, OFDM and hybrid CDMA/TDMA. However, a critical question for OFDM and hybrid CDMA/TDMA is the maturity of the technology and lack of global support. For GSM operators without new UMTS spectrum GSM384/EDGE, enhancement of the 200 kHz GSM carrier, will provide a way to offer UMTS services. In addition to these UMTS radio access technologies, wireless ATM, providing bit rates up to 20 Mbit/s, could be integrated with GSM/UMTS.

EPILOGUE

Since writing this Chapter, ETSI has made a consensus decision on the UTRA concept. The views presented in this Chapter can serve as examples of the various options that were considered during the UTRA evaluation process. The selected air interface has received a broad support from different regions participating to the GSM development. The UTRA contains following elements:

1. In the paired band (FDD - Frequency Division Duplex) of UMTS the system adopts the radio access technique formerly proposed by the WCDMA group.
2. In the unpaired band (TDD - Time Division Duplex) the UMTS system adopts the radio access technique proposed formerly by the TD-CDMA group.

The WCDMA scheme that was adopted for FDD is a harmonized version of the ETSI and ARIB schemes, which were originally based on the FMA2 and Core-A schemes, respectively. Thus, a global solution that strengthens the GSM evolution has been achieved.

REFERENCES

[1] ETSI ETR 271 Objectives and Overview of UMTS.
[2] ETSI DTR/SMG-05-0102 Vocabulary for UMTS.
[3] ETSI ETR 291 System Requirements for UMTS.
[4] ETSI ETR 312 Scenarios and considerations for the introduction of UMTS.
[5] ETSI DTR/SMG-05-0201 Framework of Services to be Supported by UMTS.
[6] ETSI DTR/SMG-05-0301 Framework of Network Architecture for UMTS.
[7] ETSI DTR/SMG-05-0401 Overall Requirements on the radio interface(s) of UMTS.
[8] ETSI DTR/SMG-05-0401 Selection Procedures for Radio Transmission principles for UMTS.
[9] ETSI DTR/SMG-05-0501 Objectives and Framework for TMN of UMTS.
[10] ETSI DTR/SMG-05-0602 Quality requirements and selection procedure for support of voice band data coding fro UMTS.
[11] ETSI DTR/SMG-05-0901 Security principles for UMTS.
[12] ETSI DTR/SMG-05-1201 Framework for satellite integration within the UMTS.
[13] ETSI DTR/SMG-05-1202 Technical characteristics, capabilities and limitations of satellite system applicable for UMTS.
[14] ETSI ETS 050-2101 System concept and reference model for the UMTS.
[15] ETSI ETS/SMG-05-02201 UMTS Service Aspects; Service principles.
[16] UMTS Forum, A Regulatory Framework for UMTS, Report no-1 from the UMTS Forum, June 1997.
[17] ERO Report on UMTS frequencies, September 1996.
[18] Andermo (ed.), "UMTS Code Division Testbed (CODIT)", CODIT Final Review Report, September 1995.
[19] A. Baier, U.-C. Fiebig, W. Granzow, W. Koch, P. Teder and J. Thielecke, "Design Study for a CDMA-Based Third Generation Mobile Radio System", IEEE Journal on Selected Areas in Communications, Vol. 12, No. 4, pp. 733 - 743, May 1994.
[20] A. Urie (ed.), "ATDMA System Definition", ATDMA deliverable R2084/AMCF/PM2/DS/R/044/b1, January 1995.
[21] A. Urie, M. Streeton and C. Mourot, "An Advanced TDMA Mobile Access System for UMTS", IEEE Personal Communications, Vol.2, No. 1, pp. 38-47, February 1995.
[22] Ojanperä Tero, K. Rikkinen, H. Hakkinen, K. Pehkonen, A. Hottinen and J. Lilleberg, "Design of a 3rd Generation Multirate CDMA System with Multiuser Detection, MUD-CDMA", Proceedings of ISSSTA96 Conference, Vol. 1, pp. 334-338, Mainz, Germany, Sep. 1996.
[23] K. Pajukoski and J.Savusalo, "Wideband CDMA Test System", Proceedings of PIMRC97, pp. 669- 672, Helsinki, September 1997.
[24] Westman Tapani and Holma Harri, "CDMA System for UMTS High Bit Rate Services", Proceedings of VTC97, pp. 824-829, Phoenix, U.S.A, May 1997.
[25] E. Nikula and E. Malkamäki, "High Bit Rate Services for UMTS using wideband TDMA carriers", Proceedings of ICUPC'96, Vol. 2, pp. 562 - 566, Cambridge, Massachusetts, September/October 1996.
[26] ETSI SMG2, "Description of Telia's OFDM based proposal" TD 180/97 ETSI SMG2, May 1997.
[27] T.Ojanperä, M. Gudmundson, P. Jung, J. Sköld, R. Pirhonen, G. Kramer and A. Toskala, "FRAMES - Hybrid Multiple Access Technology ", Proceedings of ISSSTA96 Conference, Vol. 1, pp. 320 - 324, Mainz, Germany, September 1996.

[28] T.Ojanperä, P-O. Anderson, J. Castro, L. Girard, A.Klein and R.Prasad, "A Comparative Study of Hybrid Multiple Access Schemes for UMTS", Proceedings of ACTS Mobile Summit Conference, Vol. 1, pp. 124-130, Granada, Spain, November 1996.

[29] T.Ojanperä, J. Sköld, J. Castro, L. Girard and A. Klein, "Comparison of Multiple Access Schemes for UMTS", Proceedings of VTC97, Vol.2, pp. 490-494, Phoenix, U.S.A, May 1997.

[30] A.Klein, R. Pirhonen, J. Sköld and R. Suoranta, "FRAMES Multiple Access Mode 1 - Wideband TDMA with and without Spreading" Proceedings of PIMRC97, pp. 37-41, Helsinki, September 1997.

[31] F.Ovesjö, E.Dahlman, T.Ojanperä, A.Toskala and A.Klein, "FRAMES Multiple Access Mode 2 - Wideband CDMA", Proceedings of PIMRC97, pp. 42-46, Helsinki, September 1997.

[32] T. Ojanperä, A.Klein and P.O. Anderson, "FRAMES Multiple Access for UMTS", IEE Colloquium on CDMA Techniques and Applications for Third Generation Mobile Systems, London, May 1997.

[33] ARIB FPLMTS Study Committee, "Report on FPLMTS Radio Transmission Technology SPECIAL GROUP, (Round 2 Activity Report)", Draft v.E1.1, January 1997.

[34] A. Sasaki, "A perspective of Third Generation Mobile Systems in Japan", IIR Conference Third Generation Mobile Systems, The Route Towards UMTS, London, February 1997.

[35] F. Adachi et.al., "Multimedia mobile radio access based on coherent DS-CDMA", Proc. 2nd International workshop on Mobile Multimedia Commun., A2.3, Bristol University, UK Apr. 1995.

[36] K. Ohno, M. Sawahashi and F. Adachi, "Wideband coherent DS-CDMA", Proc. IEEE VTC'95, pp. 779 -783, Chicago, U.S.A, July 1995.

[37] T. Dohi et.al., "Experiments on Coherent Multicode DS-CDMA:", Proc. IEEE VTC'96, pp. 889 - 893, Atlanta GA, USA.

[38] F.Adachi, M. Sawahashi , T. Dohi and K. Ohno, "Coherent DS-CDMA: Promising Multiple Access for Wireless Multimedia Mobile Communications", Proc. IEEE ISSSTA'96, pp. 351 - 358, Mainz, Germany, September 1996.

[39] S.Onoe, K. Ohno, K. Yamagata and T. Nakamura, "Wideband-CDMA Radio Control Techniques for Third Generation Mobile Communication Systems", Proceedings of VTC97, Vol.2, pp. 835-839, Phoenix, USA, May 1997.

[40] A. Urie, "Advanced GSM: A Long Term Future Scenario for GSM", Proceedings of Telecom 95, Vol.2 pp. 33-37, Geneva, October 1995.

[41] J. Sköld, P. Schramm, P-O. Anderson and M. Gudmundson, "Cellular Evolution into Wideband Services", Proceedings of VTC97, Vol.2, pp. 485 - 489, Phoenix, USA, May 1997.

[42] P.Ranta, A.Lappeteläinen, Z-C Honkasalo, "Interference cancellation by Joint Detection in Random Frequency Hopping TDMA Networks", Proceedings of ICUPC96 conference, Vol 1 , pp. 428-432.

[43] T. Ojanperä, "Overview of Research Activities for Third Generation Mobile Communications", in Wireless Communications TDMA vs. CDMA, S. Glisic and P. Leppanen (eds.), pp. 415-446, Kluwer Academic Publishers, 1997.

[44] A. Klein and P.W. Baier, "Linear unbiased data estimation in mobile radio systems applying CDMA", IEEE Journal on Selected Areas in Communications, vol. SAC-11, pp. 1058-1066, 1993.

[45] M.M. Naßhan, P. Jung, A. Steil and P.W. Baier, "On the effects of quantization, non-linear amplification and band limitation in CDMA mobile radio systems using joint detection", Proceedings of the Fifth Annual International Conference on Wireless Communications WIRELESS'93, pp. 173-186, Calgary/Canada, 1993.

[46] P. Jung, J.J. Blanz, M.M. Naßhan and P.W. Baier, "Simulation of the uplink of JD-CDMA mobile radio systems with coherent receiver antenna diversity", Wireless Personal Communications, An International Journal (Kluwer), vol. 1, pp. 61-89, 1994.

[47] J. Mikkonen and J. Kruys, "The Magic WAND: a wireless ATM access system", Proceedings of ACTS Mobile Summit Conference, Vol 2., pp. 535 - 542, Granada, Spain, 1996.

[48] B. Engström and C. Österberg, "A System for Test of Multi access Methods based on OFDM", Proc. IEEE VTC'94, Stockholm, Sweden, 1994.

[49] M. Ericson, H. Olofsson, M. Wahlqvist and C. Österberg, "Evaluation of the mixed service ability for competitive third generation multiple access technologies", Proceedings of VTC'97, pp. 1356-1359, Phoenix, Arizona, USA.

[50] M. Wahlqvist, R. Larsson and C. Österberg, "Time synchronization in the uplink of an OFDM system", Proc. IEEE VTC'96, pp. 1569 - 1573, Atlanta GA, USA.

[51] R. Larsson, C. Österberg and M. Wahlqvist, "Mixed Traffic in a multicarrier System", Proc. IEEE VTC'96, pp. 1259-1263, Atlanta GA, USA.

Index

D

E

F

G

H

I

J

L

M

N

O

W